Science and Ethics

In *Science and Ethics,* Bernard Rollin historically and conceptually examines the ideology that denies the relevance of ethics to science. Providing an introduction to basic ethical concepts, he discusses a variety of ethical issues that are relevant to science and how they are ignored, to the detriment of both science and society. These issues include research on human subjects, animal research, genetic engineering, biotechnology, cloning, xenotransplantation, and stem cell research. Rollin also explores the ideological agnosticism that scientists have displayed regarding subjective experience in humans and animals and its pernicious effect on pain management. Finally, he articulates the implications of the ideological denial of ethics for the practice of science itself in terms of fraud, plagiarism, and data falsification. In engaging prose and with philosophical sophistication, Rollin cogently argues in favor of making education in ethics part and parcel of scientific training.

Bernard E. Rollin is University Distinguished Professor of Philosophy, Biomedical Sciences, and Animal Sciences and University Bioethicist at Colorado State University in Fort Collins. He is the author of fourteen books, including *The Frankenstein Syndrome: Ethical and Social Issues in the Genetic Engineering of Animals* and *The Unheeded Cry: Animal Consciousness, Animal Pain and Science.*

Science and Ethics

BERNARD E. ROLLIN
Colorado State University

CAMBRIDGE UNIVERSITY PRESS
Cambridge, New York, Melbourne, Madrid, Cape Town, Singapore, São Paulo

Cambridge University Press
40 West 20th Street, New York, NY 10011-4211, USA

www.cambridge.org
Information on this title: www.cambridge.org/9780521857543

First published 2006

Printed in the United States of America

A catalog record for this publication is available from the British Library.

Library of Congress Cataloging in Publication Data

Rollin, Bernard E.
Science and ethics / Bernard E. Rollin.
p. cm.
Includes bibliographical references and index.
ISBN 0-521-85754-6 (hardback) – ISBN 0-521-67418-2 (pbk.)
1. Medicine – Research – Moral and ethical aspects. 2. Biotechnology – Moral and ethical aspects. 3. Genetic engineering – Moral and ethical aspects. 4. Human genetics – Moral and ethical aspects. 5. Human experimentation in medicine – Moral and ethical aspects. 6. Research – Moral and ethical aspects. I. Title.
[DNLM: 1. Ethics, Medical. 2. Biotechnology – ethics. W 50 R734s 2006]
R852.R67 2006
174.2′8–dc22 2005025738

ISBN-13 978-0-521-85754-3 hardback
ISBN-10 0-521-85754-6 hardback

ISBN-13 978-0-521-67418-8 paperback
ISBN-10 0-521-67418-2 paperback

To Linda and Mike

Contents

Acknowledgments *page* ix

Preface xi

1. The Waxing and Waning of Faith in Science 1
2. Scientific Ideology and "Value Free" Science 11
3. What Is Ethics? 31
4. Ethics and Research on Human Beings 66
5. Animal Research 99
6. Biotechnology and Ethics I: Is Genetic Engineering Intrinsically Wrong? 129
7. Biotechnology and Ethics II: Rampaging Monsters and Suffering Animals 155
8. Biotechnology and Ethics III: Cloning, Xenotransplantation, and Stem Cells 185
9. Pain and Ethics 215
10. Ethics in Science 247

Bibliography 275

Index 283

Acknowledgments

To my science colleagues at Colorado State University, with thanks for your friendship and collegiality, when lesser people, with lesser vision, might well have responded to my work with anger and enmity. My debt to you is incalculable. You eloquently taught me that Plato is right; thought is dialogue, people in lively discussion, not Rodin's isolated Cartesian.

Preface

In a sense, my whole career can be viewed as an attempt to articulate the legitimate role of ethics in science, on both a theoretical and a practical level. With my appointment to the Colorado State University College of Veterinary Medicine as the person charged with developing and teaching the field of veterinary medical ethics and, shortly thereafter, serving as an "ombudsman for animals" charged with achieving consensus on animal use issues in science came a unique opportunity for testing theory in practice and for almost daily interaction with scientists on ethical issues. This activity in turn meshed well with my working with colleagues in the 1970s to write legislation protecting laboratory animals, in a real way articulating the emerging social ethic for animal treatment in a manner that would benefit animals without harming research and, ideally, *improving* it by underscoring the control of hitherto ignored deforming variables resulting from uncontrolled pain and distress in animal subjects.

Ever since I was a biology student in the 1960s, I had also chafed under science teaching that ignored ethical and conceptual issues raised by biological science. Funding from the National Science Foundation in the mid-1970s allowed me, together with molecular botanist Murray Nabors, to develop a year-long, five-credit honors biology course in which ethics and philosophy were taught as part and parcel of biology. We team-taught the course for twenty-five years and were gratified when many of our students went on to become researchers, physicians, veterinarians, professors, government officials, and scientists, and would unfailingly return to thank us for the "multidimensional" view of biomedicine we had instilled.

Some ten years ago, I was asked to develop a Science and Ethics course for Ph.D. candidates in the sciences, as required by the National Institutes of Health for people receiving training grants. The course has been quite successful, albeit causing tension between my desire to keep it small yet to accommodate increasing numbers of interested science students from many fields. Interestingly enough, I found that today's students are far less willing than was my generation to accept the ideology that science was "ethics-free" and "value-free" and are much further along in this area than I was.

I also began to believe that creating ethically sophisticated scientists was a necessary condition for continued social acceptance of and support for science, a point I develop in my discussion of biotechnology.

This book is a confluence of all the aforementioned vectors. If it does not stimulate student reflection on the full range of how ethics is enmeshed in the fabric of science, as well as provoke student interest in applying ethical questioning to their own area of science, I will have failed in what I tried to do. For this reason, my style is sometimes unorthodox, mixing accounts of what I have lived through with accounts of the issues.

I am grateful to my scientist colleagues for their openness and receptivity to my gadfly role. They have treated me as a friend, giving me appointments in two science departments and the opportunity to teach in numerous science programs, and they have encouraged me to share their research and undertake my own in areas ranging from animal cognition to immunological castration of beef cattle. The same is true of my students, who are the future.

I want to thank Linda Rollin for exasperating but trenchant criticism and Michael Rollin for illuminating dialogue over twenty years.

1

The Waxing and Waning of Faith in Science

Those of us who grew up during the 1950s and early 1960s can still vividly recall the seemingly unbridled enthusiasm that society displayed toward science and technology. Sunday supplements, radio, television, and newspaper advertisements, television and radio shows, world's fairs, comic books, popular science magazines, newsreels, and, indeed, virtually all of popular culture heralded the vision of a golden age to come through science. One popular Sunday evening program sponsored by Dupont featured Ronald Reagan promising – with absolutely no irony – "better things for better living through chemistry," a slogan that evoked much hilarity during the drug-soaked 1960s.

In an age where TV dinners were symbols of modern convenience, rather than unpleasant reminders of cramped airplane trips, nothing seemed beyond the power of science. The depictions of science-based utopia – perhaps best epitomized in the Jetsons cartoons – fueled unlimited optimism that we would eventually all enjoy personal fliers, robotic servants, the conquest of disease. Expanding population? No problem – scientists would tow icebergs and desalinize water to make deserts bloom. The Green Revolution and industrialized agriculture and hydroponics would supply our nutritional needs at ever-decreasing costs. Computer gurus such as Norbert Wiener promised that cybertechnology would usher in "the human use of human beings." We would colonize the asteroids; extract gold from the sea; supply our energy needs with "clean, cheap" nuclear power; wear disposable clothing; educate our children according to sound "science-based" principles; conquer disease and repair nature's deficiencies and mistakes. "City planning," based in science, would create utopian "communities." Vestiges of this science-dependent vision have endured into the twenty-first century. Drug companies promise the design of individually targeted drugs and treatments based in biotechnology. The biotech industry still trumpets an end to famine and nutritional deficiency by way of genetic manipulation. The information technologies expand exponentially. But one no longer finds the unqualified social optimism in a science-driven future, and expressions of faith in such a future ring hollow. (Indeed, *that* vision became the stuff of numerous nostalgia coffee-table books at the advent of the millennium, perhaps because of our awareness that such a world view was born of a never-to-be-recaptured innocence and naïveté.)

One can certainly argue that our disappointments are a function of a vision too naïve and a set of unreasonable expectations regarding what science can do. Further, social reflection on increased human knowledge and its attendant control over nature well before the scientific revolution has been unrelentingly plagued by the question of whether humans have the wisdom to manage such increases in knowledge. The Tower of Babel story; the legends of Icarus and Daedalus; the Talmudic account of those rabbis who sought Cabbalistic knowledge and found only madness, apostasy, and ruination; and the story of the Golem and the Sorcerer's apprentice out of control all bespeak deep-rooted fears about advancing human knowledge and control over nature not being unequivocal goods. Indeed it is not an accident that the Bible's first moral lesson concerns the fall resulting from eating of the apple resulting in true knowledge.

With the advent of the scientific and industrial revolutions, these cautionary tales increased and intensified, with Mary Shelley's *Frankenstein* a vibrant symbol of modern concerns and, indeed, of contemporary concerns, given the endless and unabating proliferation of variations on the Frankenstein story pervading popular culture in the twentieth century.

Those reflections suggest that – even amid the most Pollyana-ish enthusiasm about science that pervaded American culture from the 1940s to the 1960s – there was a dark dimension and an ambivalence about human ability to manage proliferating knowledge and the power it conferred. And thus, even as we dreamed the Jetsonian future, we were never blinded to the strong suspicion that there could also be monsters. For this reason, our world-view of science as curer of ills and slayer of dragons was quick to shift in the face of evidence that not all was as promised.

Beginning in the 1960s, traditional American anti-intellectualism (of the sort that dismissed Adlai Stevenson's presidential candidacy because he was an "egghead") began to direct itself toward science and technology (the two have never been clearly distinguished in the American public's mind, in part because science is often promoted in terms of the technology flowing from it). "Better living through chemistry" was belied by air and water pollution. One river in Ohio – the Cuyahoga – was in fact so infused with chemical waste that it could be set on fire! People became aware that industrialization was a mixed blessing; the factories that created wealth and jobs fouled the air and water, giving flesh to William Blake's gloomy and prophetic description of them as "dark satanic mills." The automobile and the network of roads that carried it, initially the archetypes of technological blessing, became major sources of social disappointment, as cognizance of urban air pollution and traffic snarls grew. By the late 1960s, eight-lane highways and eight-lane gridlock became a favorite butt of jokes, as did the "smog" they engendered.

The growth of environmentalism in the late 1960s contributed to the demise of earlier scientific optimism. What were traditionally seen as boundless natural resources to be exploited at little cost by technology in pursuit of wealth and the science-based good life were now seen to involve hidden costs, from toxicity of air and water to loss of species and degradation of ecosystems.

The rapid growth of environmentalism, incidentally, must be viewed along with civil rights and feminism as one of the remarkably rapid and dramatic twentieth-century changes in social ethics that few anticipated. I recall a 1965 poll of 1964 graduates conducted by Phi Beta Kappa at the City College of New York, wherein the graduates were asked to rank the major problems confronting American society. Of all the hundreds of respondents, only one person listed environmental despoliation as an issue. Yet by 1969, the first massively supported Earth Day marked this major change in social gestalt, a perspective that has been enhanced, rather than diminished, by the passage of time, to the point that over 60 percent of Americans count themselves as "environmentalists," and "evil industrial polluters" have become an action movie cliché. So powerful, in fact, is the environmental mindset that it trumps even personal freedom and property rights, historically bedrock American values, as when concern about "secondhand smoke" leads to legislated antismoking bans, and concern about endangered species of any sort, not only "charismatic mega-fauna," can hold up land development (*vide* the snail darter and the Preble's jumping mouse). A rancher friend of mine was banned from haying part of his land because he might bale a jumping mouse, though none had been found on his property.

Naïve beliefs about biomedical science conquering disease and biomedical scientists as dragonslayers have given way to cynicism about the motives of scientists, drug companies, and the medical establishment and the embracing of magic-think via "alternative medicine." This disillusionment has been fueled by multiple factors: the exposure of iatrogenesis in modern medicine by critics such as Ivan Illich;[1] the failure of medicine to concern itself with quality of life and its tendency to increase life at all costs regardless of suffering; the attendant failure to control pain in the terminally ill for fear of "addiction"; the failure of the much-touted "war on cancer" to defeat cancer (though it did augment basic biological knowledge); the periodic flip-flops by the medical community on what constitutes a "healthy diet"; what I have called the "medicalization of evil," as when child abuse, youth violence, gambling, obesity, and alcoholism are labeled diseases by the medical community, a move that blatantly defies common sense; and the revelations about cavalier scientist treatment of human and animal research subjects. These have collectively eroded the view of biomedicine as a moral science, and set what we shall shortly call the common sense of science at loggerheads with ordinary common sense.

One highly touted techno-scientific advance was the so-called green revolution: the attempt to increase crop yield by use of scientific principles. A parallel movement in animal agriculture led to the change in that field from seeing itself as based in animal husbandry – care for animals – and instead as animal science – defined in textbooks as "the application of industrial methods to the production of animals." These congruent developments, initially met with public enthusiasm,

[1] See Illich, *Medical Nemesis.*

have in fact become identified in the public mind with generating Frankensteinean results from scientists' hubris. Modern agriculture is now widely seen as being based in avaricious petrochemical consumption and thus as not "sustainable"; as being thereby a major cause of air and particularly water pollution; as relying on economies of scale that lead inexorably to corporate domination of agriculture and to the loss of family farms and rural communities; as degrading farm labor; as putting small operators and farm workers out of business; as eroding food quality and increasing dangers coming from the food supply by reliance on herbicides, pesticides, hormones, and antibiotics; as depleting the land and hurting the animals; and as generating monoculture.

At the same time, public confidence in scientific reassurances has precipitously diminished as a result of an apparently endless list of scientific prognostications gone afoul. The escape of "killer" bees, the *Challenger* disaster, Three Mile Island and Chernobyl, blackouts and brownouts, manipulation of scientific data by cigarette companies, thalidomide, Fen-Phen, Vioxx, the University of Pennsylvania head-injury videotapes of baboon abuse, and the well-publicized cases of people hurt and killed in research have all diminished our faith in "trust me – I'm a scientist" and nurtured the resurgence of the Frankenstein view of scientist as dangerous, whether through misguided good intentions (Dr. Frankenstein's intentions were to augment life), incompetence, corruption, or simply biting off more than he or she can chew.

Another factor associated with diminished confidence in science is the advent – or resurgence – of a mystical streak in society. (I use the phrase "associated with" because it is difficult to tell whether the draw of the occult is a cause or an effect of diminished faith in science, or perhaps both cause and effect.) The key point is that, for whatever reason, beliefs inimical to a skepticism forged in science have reached epidemic proportions. Thousands of educated women now affirm a belief in Wicca, the primary manifestation of witchcraft, allegedly an ancient body of wisdom suppressed by male domination. Millions pursue astrology, unfazed by either its predictive failures or its vacuity ("Your life will change"). Millions of others sport crystals or minerals for their "positive energy." Most impressively, "alternative" medicine and alternative veterinary medicine are thriving – according to the American Medical Association, in one year the U.S. public spent

$29 billion on such unproven therapies whose efficacy, safety, and batch consistency remains unproven and usually untested. It seems that if a putative treatment modality comes from Asia, it is particularly valued – witness the huge success of acupuncture, acupressure, and Reiki. Treatments that violate all known laws of science flourish anyway; witness the resurgence of homeopathy or Bach flower essences, where substances are diluted to the point where they are chemically incapable of any biological activity, or the "healing touch." Others, such as magnet therapy, flourish despite having been demonstrated to show no effect.

Cults, sects, and hermetic traditions are a growth industry, as are books on allegedly magic texts of the "The Kabbalah and You" ilk. Perhaps most astounding is the resurgence of exorcism among both Catholics and Protestants, as well as among some psychiatrists, who admonish all of us to mark the difference between mental and behavioral problems that represent genuine illnesses, versus the easily mislabeled cases of *demonic possession* with which mental illnesses may be confused![2]

In my mind, however, the most critical factor leading to social disenchantment with science has been the singular failure of the scientific community to engage the myriad ethical issues emerging from scientific activity. This is particularly problematic in an age that is suffused with ethical concern, a situation that paradigmatically characterizes the United States during the last half-century.

There is an ancient curse that is most appropriate to the society in which we live: "May you live in interesting times." From the point of view of our social ethics, we do indeed live in bewildering and rapidly changing times. The traditional, widely shared, social ethical truisms that gave us stability, order, and predictability in society for many generations are being widely challenged by women, ethnic minorities, homosexuals, the handicapped, animal rights advocates, internationalists, environmentalists, and more. Most veterinarians now realize, to take a very obvious example, that society is in the process of changing its view of animals and our obligations to animals. Laboratory animal veterinarians have probably seen the most clearly articulated evidence of such a changing ethic, but it is also patent to any companion

[2] Cuneo, *American Exorcism.*

animal practitioners, food animal practitioners, or zoo veterinarians who take the trouble to reflect on the new social expectations shaping and constraining the way they do their jobs.

It is very likely that there has been more and deeper socio-ethical change since the middle of the twentieth century than has occurred during centuries of an ethically monolithic period such as the Middle Ages. Anyone over forty has lived through a variety of major moral earthquakes; the sexual revolution, the end of socially sanctioned racism, the banishing of IQ differentiation, the rise of homosexual militancy, the end of "*loco parentis*" in universities, the advent of consumer advocacy, the end of mandatory retirement age, the mass acceptance of environmentalism, the growth of a "sue the bastards" mind-set, the implementation of affirmative-action programs, the rise of massive drug use, the designation of alcoholism and child abuse as diseases rather than moral vices, the rise of militant feminism, the emergence of sexual harassment as a major social concern, the demands by the handicapped for equal access, the rise of public suspicion of science and technology, the mass questioning of animal use in science and industry, the end of colonialism, and the rise of political correctness all are examples of the magnitude of ethical change during this brief period.

With such rapid change come instability and bewilderment. Do I hold doors open for women? (I was brought up to do so out of politeness, but is such an act patronizing and demeaning?) Do I support black student demands for black dormitories (after I marched in the 1960s to end segregation)? Am I a bad person if I do not wish to hire a transsexual? Can I criticize the people of Rwanda and Bosnia for the bloodbaths they conduct without being accused of insensitivity to cultural diversity? Do I obey the old rules or the new rules? Paradoxically, the appeal to ethics and the demand for ethical accountability have probably never been stronger and more prominent – witness the forceful assertion of rights by and for people, animals, and nature – yet an understanding of ethics has never been more tentative, and violations of ethics and their attendant scandals in business, science, government, and the professions have never been more prominent. There is probably more talk of ethics than ever – more endowed chairs, seminars, conferences, college courses, books, media coverage, journals devoted to ethical matters than ever before – and yet, ironically, most people probably believe that they understand ethics far less than their progenitors did. Commonality of values has given way to plurality and diversity; traditions are being eroded; even the church is no longer the staunch defender of traditional ethical norms.

Thus ethics is in the air; "ethics sells," as one textbook salesman crassly put it to me. "Applied ethics" courses, virtually nonexistent in the 1960s, are a growth industry and saved many philosophy departments during the mercenary 1980s. Indeed, the rise of medical ethics, and particularly of medical ethics "think-tanks," was, at least in part, a self-defense move to protect the medical community. Historically accustomed to not being questioned, the medical community found itself dealing with a public that, thanks to television and other media coverage, was fairly well versed in issues of medical ethics.

Unfortunately, medical ethics, which in my view has been very establishment-oriented and tame, must still be seen as exhibiting moral sophistication compared with science in general. (One of my friends, a pioneer in medical ethics in the 1970s, explained bitterly that medical ethicists tamely focused on "high visibility" issues such as pulling the plug on the irreversibly comatose Karen Ann Quinlan, while totally avoiding the far more important issue of fee for service.) For, by and large, the research community has failed abysmally to engage virtually any ethical questions flowing from its activities. For example, issues that were manifest to the general public in biomedical research – invasive and abusive use of human and animal subjects – were essentially invisible to the research community. One can search scientific journals, conferences, textbooks, and the like and find almost no solid discussions of the ethical issues raised by experimentation. When the research community did finally engage the question of animal research in the early 1980s, upon its realization that much-dreaded legislation was a real threat, it did so in a highly emotive way that was in fact not that far from the style utilized by its antivivisectionist opponents, with frankly outrageous claims that any constraints on animal use would unequivocally forestall medical progress and harm the health of children. This was in turn a reflection of the view that ethical issues can be approached only emotionally, never rationally, which was rife in the scientific community.

We shall elaborate on these issues and the mentality that led to their mishandling as we proceed through our discussion. For now, it suffices to point out that the research community's mind-set on ethics is still largely unchanged, despite the lessons that should have been learned from the animal experimentation issues in the 1980s. The area of biotechnology provides a profound – and troubling – current example of the way in which the scientific community fails to engage ethical issues, which in turn leads to public rejection of the science or technology in question, for bad reasons. This has occurred with genetic engineering, genetically modified foods in Europe, cloning, and stem cell technology. This, in turn, gives further evidence that willful ignoring of ethical issues is one of the major reasons for public disenchantment with science.

Any new technology will create a lacuna in socio-ethical thought, and the newer and more powerful the technology, the greater the vacuum. Will a given technology improve our lives or degrade them? In what ways? Which aspects of the technology need to be controlled, regulated, accepted, or rejected to assure that it is a force for good, not for ill? Will it erode or enhance our autonomy? So it is surely incumbent upon those who develop a technology and best understand its strengths and limitations to help society think such issues through. If they fail to do so, the ethical implications vacuum may be filled by doomsayers: political, religious, or other vested interests who may totally distort, exaggerate, or minimize the issues occasioned by the technology and induce in society fear that leads to irrational rejection of the technology or to naïve enthusiasm that leads to imprudent acceptance of it.

This is exactly what happened with biotechnology, leading to its summary rejection in Europe and to lesser but significant social concern in the United States. The research community totally failed to articulate the ethical implications of cloning, genetic engineering, genetically modifying food, BST (bovine somatotropin) use in cattle, developing biomedical animal models for human genetic diseases, and so on, leaving a vacuum in social thought. Religious leaders and apocalyptic doomsayers such as Jeremy Rifkin immediately filled that lacuna with worst case but meaningless slogans – genetic engineering is against God, cloning is against nature, biotechnology has man "playing God" or usurping his role, and so on, illustrating what I have called a Gresham's law for ethics: bad ethics driving good ethics out of circulation, analogous to Gresham's realization that "bad money" in circulation (e.g., valueless paper deutsche marks) leads to hoarding of "good money" (e.g., gold). No one will pay a debt with gold if they can pay with near-valueless currency.

Research funding was displaced by public fear; laws were quickly passed against cloning. Leaders of the regulatory community steadfastly refused to mandate labeling of GM foods, affirming that they do not differ from normal foods save in the "process" of formation – the product is the same. No one discussed ethics rationally, since the research community tends to believe that one cannot do so, and the other side didn't try to – it was doing fine with sloganeering. Regulators strongly downplayed the risks of biotechnology while ignoring excellent research showing that ethics was of far greater concern (at least to the European public) than risk. The net effect? Substantial portions of the European Community are dead set – and powerfully – against genetically modified foods, and the U.S. public cannot yet see the enormous power for good potentially inherent in biotechnology, the most powerful technology ever devised. Even Monsanto, which spent a fortune on developing and marketing BST for increasing milk production, failed to consider the ethical dimensions of the technology as perceived by final milk consumers, rather than by producers. In our discussion below, we explore many of these neglected ethical issues in depth. If we do not produce a generation of scientists who can think in ethical terms and lead public ethical discussions of science, we may lose countless real benefits of scientific advances, as well as public support of science.

2

Scientific Ideology and "Value Free" Science

Before exploring specific ethical issues that the scientific community has mishandled or failed to handle, we must first address a basic question: Why does the research community have such a bad track record in dealing with ethics? Why has it consistently missed the mark set by society for rational ethical discussion and explanation? And what should it be doing instead? In my view, the problem grows out of strongly and unquestioningly held beliefs in the scientific community about science and ethics, beliefs that are never questioned to the extent that they constitute a hardened and unshakeable ideology that I have called "scientific common sense" or "scientific ideology," which stands in the same relationship to scientists' thinking that ordinary common sense does to the thinking of nonscientists. It is to this ideology we now turn.

What is an ideology? In simple terms, an ideology is a set of fundamental beliefs, commitments, value judgments, and principles that determine the way someone embracing those beliefs looks at the world, understands the world, and is directed to behave toward others in the world. When we refer to a set of beliefs as an ideology, we usually mean that, for the person or group entertaining those beliefs, nothing counts as a good reason for revising those beliefs, and, correlatively, raising questions critical of those beliefs is excluded dogmatically by the belief system. (As David Braybrooke has stated it, "ideologies distort as much by omitting to question as by affirming answers.")[1]

[1] David Braybrooke, "Ideology" in *Encyclopedia of Philosophy*, vol. 2, p. 126.

The term is most famously, perhaps, associated with Marx, who described capitalist ideology (or free market ideology) as involving the unshakeable beliefs that the laws of the competitive market are natural, universal, and impersonal; that private property in ownership of means of production is natural, permanent, and necessary; that workers are paid all they can be paid; and that surplus value should accrue to those who own the means of production.

Though most famously associated with the Marxist critique of capitalism, we all encounter ideologies on a regular basis. Most commonly, perhaps, we meet people infused with religious ideologies, such as biblical fundamentalism, who profess to believe literally in the Bible as the word of God. I have often countered such people by asking them whether they have read the Bible in Hebrew and Greek, for surely God did not speak in antiquity in English. Further, I point out, if they have not read the original language, they are relying on interpretations rather than literal meaning, since all translation *is* interpretation, and interpretation may be wrong. To illustrate this point, I ask them to name some of the Ten Commandments. Invariably, they say, "Thou shalt not kill." I then point out that the Hebrew in fact does not say, "Thou shalt not kill"; it says "Thou shalt not *murder*." This then should be enough to convince them that they do not believe the Bible literally, if only because they cannot read it literally. Does it do so? Of course not. They have endless ploys to avoid admitting that they can't possibly believe it literally, for example, "The translators were Divinely inspired," and so on.

We of course are steeped in political ideology in grade school and high school, for example, on issues of "human equality." Ask the average college student (as I have done many times) what is the basis for professing equality, when people are clearly unequal in brains, talent, wealth, athletic ability, and so on. Few will deny this, but most will continue to insist on "equality" without any notion that "equality" refers to a way we believe we ought to treat people, not to a factual claim. If they do see equality as an "ought" claim, almost none can then provide a defense of why we believe we ought to treat people equally if in fact they are not equal. And so on. But virtually never will such a student renounce the belief in equality.

Of late, students have been steeped in the ideology of diversity and multiculturalism, affirming that no culture is superior to any other, and an admixture of cultures is always best. Few can respond to the query I tender: "What? Are you telling me that a culture where clitorectomies are performed without consent or anesthesia on helpless female children is as good as a culture that disavows such mutilations?"

Similarly, surely no one would argue that the Taliban culture, wherein women were not allowed to be educated and were beaten for laughing in public, and men were beaten for flying kites or listening to music, is as good a culture as ours. Similarly, I point out that the price of diversity is often friction and tension. No sane New Yorker leaves his or her apartment unlocked; in rural Wyoming that is de rigueur: People share common grazing land and someone may be rounding up cattle when a storm strikes, so everyone leaves their ranches unlocked in case someone needs refuge. A person in trouble is expected to enter the empty home, use the bed, make a meal, tidy up, and leave. In return, one does the same thing for others. Similarly, if one has an accident or car trouble in Wyoming, everyone stops to help. In my view, this is made possible by virtue of the fact that the culture is monolithic rather than diverse, and everyone shares the same values, beliefs, and expectations.

Thus, despite one's ability to provide cases where ethnic multiplicity or diversity have downsides, and other cases where common sense shows that some cultures are worse than others, students who have been ideologically brainwashed simply filter out such arguments, even as Marxists filter out and ignore counterexamples to their basic ideology, and fundamentalists do the same.

Ideologies are attractive to people; they give pat answers to difficult questions. It is far easier to give an ingrained response than to think through each new situation. Militant Muslim ideology, for example, sees Western culture as inherently evil and corruptive of Islam; the United States as "the Great Satan" and fountainhead of Western culture that in turn is aimed at destroying Islamic purity. The United States is thus automatically wrong in any dispute, and any measures are justified against that country in the ultimate battle against defilement.

What is wrong with ideology, of course, is precisely that it truncates thought, providing simple answers and, as Braybooke indicated in the passage quoted earlier, cutting off certain key questions. Intellectual subtlety and the powerful tool of reason, making distinctions, are totally lost to gross oversimplifications. Counterexamples are ignored. I recall working in a warehouse where the preponderance of blue-collar employees was strongly possessed of racist ideology, particularly antiblack ideology. It was universally believed that blacks were lazy, unintelligent, sneaky, crooked. One day I had an inspiration. There was in fact one African American (Joe) who worked in the warehouse and was well liked. I raised this counterexample with them. "Surely," I said, "this case refutes your claim about *all* black people." "Not at all," they said. "Joe is different – he hangs around with us."

But it is not only that ideology constricts thought. It can also create monsters out of ordinary people by overriding common sense and common decency. We have seen this manifested plainly throughout the history of the twentieth century. The recent experiences of Eastern Europe and Africa make manifest that ideologically based hatreds, whose origins have been obscured by the passage of time, may, like anthrax spores, reemerge as virulent and lethal as ever, unweakened by years of dormancy. Most strikingly, perhaps, the work of historian Daniel Goldhagen has demonstrated the enormous power of ideology to overwhelm and obscure both common sense and common decency, even among the most civilized of people.

In his monumental work, *Hitler's Willing Executioners* (1996), Goldhagen has shown that under the Nazis, ordinary Germans willingly and voluntarily engaged in genocidal activities, even when it was patently open to them to refuse to do so without fear of recriminations. The killers studied by Goldhagen were neither sadists and psychopaths of the sort attracted to the SS nor the sort of street brawlers and bullies that composed the ranks of Ernst Röhm's SA. Rather, they were normal, largely nonviolent family men, who operated neither out of fear of punishment for disobedience (one standard explanation) nor out of the blind obedience suggested by Stanley Milgram[2] and often invoked to explain Nazi killing. According to Goldhagen, neighbors became killers because of their immersion in two centuries of ideological dogma depicting Jews as pathogens in the body politic, rendering that body ill and infirm and demanding radical excision of the disease-causing organisms. As absurd as this seems to those of us unsteeped in similar ideology, it was common sense to Goldhagen's Germans and

[2] Milgram, *Obedience to Authority*.

a straightforward justification for actions they would recoil from in nonideological contexts.

As we have seen, ideologies operate in many different areas: religious, political, sociological, economic, ethnic. Thus it is not surprising that an ideology would emerge with regard to science, which is, after all, the dominant way of knowing about the world in Western societies since the Renaissance.

Indeed, knowing has had a special place in the world since antiquity. Among the pre-Socratics – or *physikoi* as Aristotle called them – one sometimes needed to subordinate one's life unquestioningly to the precepts of a society of knowers, as was the case with the Pythagoreans. And the very first line of Aristotle's *Metaphysics* – or First Philosophy – is "All men by nature desire to know." Thus the very *telos* of humanity, the "humanness" of humans, consists in exercising the cognitive functions that separate humans from all creation. Inevitably, the great knowers, such as Aristotle, Bacon, Newton, and Einstein, felt it necessary to articulate what separated legitimate empirical knowledge from spurious knowledge and jealously to guard and defend that methodology from encroachment by false pretenders to knowledge.

Thus the ideology underlying modern (i.e., postmedieval) science has grown and evolved along with science itself. And a major – perhaps *the* major – component of that ideology is a strong positivistic tendency, still regnant today, of believing that real science must be based in experience, since the tribunal of experience is the objective, universal judge of what is really happening in the world.

If one asks most working scientists what separates science from religion, speculative metaphysics, or shamanistic world views, they would unhesitatingly reply that it is an emphasis on validating all claims through sense experience, observation, or experimental manipulation. This component of scientific ideology can be traced directly back to Newton, who proclaimed that he did not "feign hypotheses" ("*hypotheses non fingo*") but operated directly from experiences. (The fact that Newton in fact *did* operate with nonobservable notions such as gravity or, more generally, action at a distance did not stop him from ideological proclamations affirming that one should not do so.) The Royal Society members apparently took him literally, went around gathering data for their commonplace books, and fully expected major scientific breakthroughs to emerge therefrom. (This idea of truth revealing itself through data gathering is prominent in Francis Bacon.)

The insistence on experience as the bedrock for science continues from Newton to the twentieth century, where it reaches its most philosophical articulation in the reductive movement known as logical positivism, a movement that was designed to excise the unverifiable from science and, in some of its forms, formally to axiomatize science so that its derivation from observations was transparent. A classic and profound example of the purpose of the excisive dimension of positivism can be found in Einstein's rejection of Newton's concepts of absolute space and time, on the grounds that such talk was untestable. Other examples of positivist targets were Bergson's (and other biologists') talk of life force (*élan vital*) as separating the living from the nonliving or the embryologist Driesch's postulation of "entelechies" to explain regeneration in starfish.

Although logical positivism took many subtly different and variegated forms, the message, as received by working scientists and passed on to students (including myself), was that proper science ought not to allow unverifiable statements. This was no doubt potentiated by the fact that the British logical positivist A. J. Ayer wrote a book that was relatively readable, vastly popular (for a philosophy book), and aggressively polemical that defended logical positivism. Entitled *Language, Truth, and Logic*; it first appeared in 1936 and has remained in print ever since.[3] Easy to read, highly critical of wool-gathering, speculative metaphysics and other soft and ungrounded ways of knowing, the book was long used in introductory philosophy courses and, in many cases, represented the only contact with philosophy that aspiring young scientists – or even senior scientists – enjoyed.

Be that as it may, the positivist demand for empirical verification of all meaningful claims became a mainstay of scientific ideology from the time of Einstein to the present. Insofar as scientists thought at all in philosophical terms about what they were doing, they embraced the simple but to them satisfying positivism we have described. Through it, one could clearly, in good conscience, dismiss religious claims, metaphysical claims, or other speculative assertions not merely as false and irrelevant to science but as meaningless. Only what could *in principle* be

[3] Ayer, *Language, Truth, and Logic*.

verified (or falsified) empirically was meaningful. "In principle" meant "someday," given technological progress. Thus, though the statement "There are intelligent inhabitants on Mars" could not in fact be verified or falsified in 1940, it was still meaningful, since we could see how it could be verified, that is, by building rocket ships and going to Mars to look. Such a statement stands in sharp contradiction to the statement "There are intelligent beings in Heaven," because, however our technology is perfected, we don't even know what it would be like to visit Heaven, it not being a physical place.

What does all this have to do with ethics? Quite a bit, it turns out. The philosopher Ludwig Wittgenstein, who greatly influenced the logical positivists, once remarked that if you take an inventory of all the *facts* in the universe, you will not find it a *fact* that killing is wrong. In other words, ethics is not part of the furniture of the scientific universe. You cannot, in principle, test the proposition that "killing is wrong." It can neither be verified nor falsified. So, empirically and scientifically, ethical judgments are meaningless. From this, it was concluded that ethics is outside the scope of science, as are all judgments regarding values rather than facts. The slogan that I in fact learned in my science courses in the 1960s, and which has persisted to the present, is that "science is value-free" in general, and "ethics-free" in particular.

This denial in particular of the relevance of ethics to science was taught both explicitly and implicitly. One could find it explicitly stated in science textbooks. For example, in the late 1980s when I was researching a book on animal pain, I looked at basic biology texts, two of which a colleague and I actually used, ironically enough, in a honors biology course we team-taught for twenty-five years attempting to combine biology and the philosophical and ethical issues it presupposed and gave rise to. The widely used Keeton and Gould textbook *Biological Science*, for example, in what one of my colleagues calls the "throat-clearing introduction," wherein the authors pay lip service to scientific method, a bit of history, and other "soft" issues before getting down to the parts of a cell and the Krebs cycle, loudly declares that "science cannot make value judgments ... cannot make moral judgments." In the same vein, Mader,[4] in her popular biology text, asserts that "science does not make ethical or moral decisions." The standard

[4] Mader, *Biology: Evolution, Diversity, and the Environment.*

line affirms that science at most provides society with *facts* relevant to making moral decisions, but never itself makes such decisions.

In addition to being explicitly affirmed, this component of scientific ideology was implicitly taught in countless ways. For example, student moral compunctions about killing or hurting an animal, whether in secondary school, college, graduate school, or professional school, were never seriously addressed until the mid- to late 1980s, when the legal system began to entertain conscientious objections. One colleague of mine, in graduate school in the late 1950s studying experimental psychology, tells of being taught to "euthanize" rats after experiments by swinging them around and dashing their heads on the edge of a bench to break their necks. When he objected to this practice, he was darkly told, "Perhaps you are not suited to be a psychologist." In 1980, when I began to teach in a veterinary school, I learned that the first laboratory exercise required of the students the third week of their first year was to feed cream to a cat and then, using ketamine (which is not an effective analgesic for visceral pain but instead serves to restrain the animal), do exploratory abdominal surgery ostensibly to see the transport of the cream through the intestinal villi. When I asked the teacher what was the point of this horrifying experience (the animals vocalized and showed other signs of pain), he told me that it was designed to "teach the students that they are in veterinary school, and needed to be tough, and that if they were 'soft,' to 'get the hell out early.'"

As late as the mid-1980s, most veterinary and human medical schools *required* that the students participate in bleeding out a dog until it died of hemorrhagic shock. Although Colorado State University's veterinary school abolished the lab in the early 1980s for ethical reasons, the department head who abolished it after moving to another university was *defending* the same practice ten years later, and explained to me that if he didn't, his faculty would force him out. As late as the mid-1990s, a medical school official told my veterinary dean that his faculty was "firmly convinced" that one could not "be a good physician unless one first killed a dog." In his autobiographical book *Gentle Vengeance*,[5] which deals with an older student going through Harvard

[5] Lebaron, *Gentle Vengeance.*

Medical School, the author remarks in passing that the only purpose he and his peers could see to the dog labs was to assure the students' divestiture of any shred of compassion that might have survived their premedical studies.

Surgery teaching well into the 1980s was also designed to suppress compassionate and moral impulses. In most veterinary schools, animals were utilized repeatedly, from a minimum of eight successive survival surgeries over two weeks to over twenty times at some institutions. This was done to save money on animals, and the ethical dimensions of the practice were never discussed, nor did the students dare raise them.

At one veterinary school, a senior class provided each student with a dog, and the student was required to do a whole semester of surgery on the animal. One student anesthetized the animal, beat on it randomly with a sledge hammer, and spent the semester repairing the damage. He received an "A."

The point is that these labs in part taught students not to raise ethical questions and that ordinary ethical concerns were to be shunted aside and ignored in a scientific or medical context. So the explicit denial of ethics in science was buttressed and taught implicitly in practice. If one did raise ethical questions, they were met with threats or a curt, "This is not a matter of ethics, but of scientific necessity," a point we see repeated when discussing questionable research on human beings.

Even at the height of concern about animal use in the 1980s, scientific journals and conferences did not rationally engage the ethical issues occasioned by animal research. It was as if such issues, however much a matter of social concern, were *invisible* to scientists, which in a real sense they in fact were. One striking example is provided by a speech given by James Wyngaarden, Director of the National Institutes of Health in 1989. The NIH director is arguably the chief biomedical scientist in the United States and certainly is a symbol of the research establishment. Wyngaarden, an alumnus of Michigan State University, was speaking to a student group at his alma mater and was asked about ethical issues occasioned by genetic engineering. His response was astonishing to lay people, though perfectly understandable given what we have discussed about scientific ideology. While new areas of science are always controversial, he opined, "science should not be hampered by ethical considerations."[6] Probably no other single incident shows as clearly the denial of ethics in science. When I read the unattributed quotation to my students and ask them to guess its author, they invariably respond "Adolf Hitler."

Nor is this sort of response restricted to biomedicine. Some years ago, PBS ran a documentary special on the Manhattan Project, which developed the atomic bomb. Scientists on the project were asked about the ethical dimensions of their work. They replied that the ethics was not their business; society makes ethical decisions, scientists simply provide technical expertise regarding the implementation of those decisions. In fact, every time I am interviewed by a reporter on ethical issues in science, my raising the "science is value-free" component of scientific ideology elicits a shock of recognition. "Oh yeah," they say, "scientists always say that when we ask them about controversial issues like weapons development."

We have argued that the logical positivism that informed scientific ideology's rejection of the legitimacy of ethics dismissed moral discussion as empirically meaningless. That is not, however, the whole story. Positivist thinkers felt compelled to explain why intelligent people continued to make moral judgments and continued to argue about them. They explained the former by saying that when people make assertions such as "killing is wrong," which seem to be statements about reality, they are in fact describing nothing. Rather, they are "emoting," expressing their own revulsion at killing. "Killing is wrong" really *expresses* "Killing, yuck!" rather than describing some state of affairs. And when we seem to debate about killing, we are not really arguing *ethics* (which one can't do any more than you and I can debate whether we like or don't like pepperoni), but rather disputing each other's *facts*. So a debate over the alleged morality of capital punishment is my expressing revulsion at capital punishment while you express approval, and any debate we can engender is over such factual questions as whether or not capital punishment serves as a deterrent against murder.

It is therefore not surprising that when scientists are drawn into social discussions of ethical issues they are every bit as emotional as their untutored opponents. (We shall illustrate this clearly in our

[6] Michigan State *News* (February 27, 1989), p. 8.

chapter on animal research.) It is because their ideology dictates that these issues *are nothing but emotional,* that the notion of rational ethics is an oxymoron, and that he who generates the most effective emotional response "wins." So, for example, in the '70s' and '80s' debate on the morality of animal research, scientists either totally ignored the issue or countered criticisms with emotional appeals to the health of children. For example, in one film entitled *Will I Be All Right, Doctor?* (the question asked by a frightened child of a pediatrician), the response was, "Yes, if *they* leave us alone to do what we want with animals." So appallingly and unabashedly emotional and mawkish was the film that when it was premiered at the American Association for Laboratory Animal Science (AALAS) meetings, a putatively sympathetic audience, the only comment forthcoming from the audience came from a veterinarian, who affirmed that he was "ashamed to be associated with a film that is pitched lower than the worst anti-vivisectionist clap-trap!"

Just how extraordinarily incapable scientists were of responding to rational ethical argument was driven home to me when I ran a long session on animal ethics and legislation at another AALAS national meeting, where I carefully laid out the arguments for legislating protections for research animals. Though the audience of laboratory animal veterinarians expressed great frustration that researchers did not listen to them, particularly in human medical schools, and that their expertise, if attended to, would make for better animal care and better science, they steadfastly refused to support their own legislative empowerment, since they opposed the importation of ethics into science.

As irrational as that was, it paled in comparison to what occurred after my session. Reporters converged on the president of AALAS, asking him to comment on my demand for legislated protection for animals. "Oh, that is clearly wrong," he said. "Why?" they queried. "Because God said we could do whatever we wish with animals," he affirmed. The reporters then turned to me and asked me to respond. Amazed that the head of a scientific organization could so invoke the Deity with a straight face (imagine the head of the American Physical Society responding to budget cuts in the funding of physics by saying, "God said we must fund physics"), I poked fun at his reply. "I doubt he is correct," I answered. "He comes from Kansas State University."

"So what?" said the reporters. "Simple," I replied. "If God chose to reveal his will at a veterinary school, it certainly would not be at Kansas! It would be at Colorado State, which is God's country!"

What are we to say of the aspect of scientific ideology that denies the relevance of values in general and ethics in particular to science?

As I hope the astute reader has begun to realize, as a human activity, embedded in a context of culture, and addressed to real human problems, science cannot possibly be value-free, or even ethics-free. As soon as scientists affirm that controlled experiments are a *better* source of knowledge than anecdotes; that double-blind clinical trials provide better proof of hypotheses than asking the Magic 8-Ball; or, for that matter, that science is a better route to knowledge of reality than mysticism, we encounter value judgments as presuppositional to science. To be sure, they are not ethical value judgments, but rather *epistemic* ("pertaining to knowing") ones, but they are still enough to show that science does depend on value judgments. So choice of scientific method or approach represents a matter of value. Scientists often forget this obvious point; as one scientist said to me, "We don't make value judgments in science; all we care about is knowledge."

In fact, reflection on the epistemic basis of science quickly leads to the conclusion that this basis includes moral judgments as well. Most biomedical scientists will affirm that contemporary biomedicine is logically (or at least practically) dependent on the sometimes invasive use of animals, as the only way to find out about fundamental biological processes. Every time one uses an animal in an invasive way, however, one is making an implicit moral decision, namely, that the information gained in such use morally outweighs the pain, suffering, distress, or death imposed on such an animal to gain the knowledge or that it is morally correct to garner such knowledge despite the harm done to animals. Obviously, most scientists would acquiesce to that claim, but that is irrelevant to the fact that it is still a moral claim.

Exactly the same point holds regarding the use of human beings in research. Clearly, unwanted children or disenfranchised humans are far better (i.e., more "high fidelity") models for the rest of us than are the typical animal models, usually rodents. We do not, however, allow unrestricted invasive use of humans despite their scientific superiority. Thus another moral judgment is presuppositional to biomedical science.

I was once arguing with a scientist colleague about the presence of moral judgments in science. He was arguing their absence. I invoked the argument that, if science was ethics-free, we would always use the highest fidelity model in our researches, thus deploying unwanted children rather than rats. In the ensuing silence, I asked him again, "Why not use the children?" "Because they won't let us!" he snapped.

In any case, many other valuational and ethical judgments appear in science, not just those involved in methodology. Which subjects and problems scientists are funded to pursue – AIDS, nonpolluting energy sources, alcoholism, but not the tensile strength of blonde hair or the intelligence of frogs – depends on social value judgments, including ethical ones. The once popular scientific subject of race or the measurement of an alleged biological property called IQ are now forbidden subjects for ethical reasons, as are myriad other subjects inimical to current socio-ethical dogmas and trends.

Even experimental design in science is constrained by ethical value judgments. The statistical design of an experiment testing the safety of a human drug will invariably deploy far greater statistical stringency than a similar experiment testing the safety of an animal drug used for precisely the same disease in animals, for ethical reasons of valuing harm to people as a much greater moral concern than harm to animals.

The root paired concepts of biomedical science – health and disease – can also be readily shown to contain irrevocably valuational components. Physicians are convinced that the judgment that something is diseased or sick is as much a matter of fact as is the judgment that the organism is bigger or smaller than a breadbox. Diseases are repeatable entities to be scientifically discovered, and physicians are scientists. This scientific stance has been repeatedly noted in its nonsubtle manifestations; anyone who has been in a hospital is aware of the tendency of physicians to see patients as instances of a disease rather than as unique individuals – science, after all, deals with the repeatable and lawlike aspects of things, not with individuals qua individuals. This tendency to remove individuality is a chronic complaint of patients – it is demeaning to be treated as an instance of something. Indeed, it is less often noticed that this tendency is medically pernicious as well. When it comes to dispensing pain medication, for example, it has been shown that pain-tolerance thresholds (i.e., the maximum pain a person can tolerate) differ dramatically across individuals and that thresholds can be modulated by a variety of factors, not the least of which is surely rapport with the physician, or the sense that the physician cares about the patient's pain. Among physicians, only Oliver Sacks, in *Awakenings,* has stressed the extraordinary degree to which a disease varies with the individual, in all of his or her complexity.

Ordinary common sense (but not medical common sense) recognizes this much. The more subtle sense in which scientism in its emphasis on fact versus value – with only the former term entering into the medical situation – misses the mark is in its understanding of the very nature of disease. For the concept of a disease, of a physical (or mental) condition in need of fixing, is inextricably bound up with valuational presuppositions. Consider the obvious fact that the concept of disease is a concept that, like good and bad, light and dark, acquires its meaning by contrast with its complement, in this case, the concept of health. One cannot have a concept of disease without at least implicit reference to the concept of health (= okay and not in need of fixing). Yet the concept of health clearly makes tacit or explicit reference to an ideal for the person or other organism; a healthy person is one who is functioning as we believe people should. This ideal is clearly valuational; most of us do not feel that people are healthy if they are in constant pain, even though they can eat, sleep, reproduce, and so forth. That is because our ideal for a human life is really an ideal for a good human life – in all of its complexity.

Health is not merely what is statistically normal in a population (statistical normalcy can entail being diseased), nor is it purely a biological matter. The World Health Organization captures this idea in its famous definition of health as "a complete state of mental, physical, and social well-being." In other words, health is not just of the body. Indeed, the valuational dimension is both explicit and not well defined, for what is "well-being" save a value notion to be made explicit in a socio-cultural context?

Heedless of this point, and wedded to the notion that disease is discovered by reference to facts, not in part decided by reference to values, physicians make decisions that they think are discoveries. When physicians announce that obesity is the number one disease in the United States, and this "discovery" makes the cover of *Time* magazine, few people, physicians or otherwise, analyze the deep structure of that statement. Are fat people really sick people? Why? Presumably, the physicians who make this claim are thinking of something like this: Fat people tend to get sick more often – flat feet, strokes, bad backs, heart conditions. But, one might ask, is something that makes you sick itself a sickness? Boxing may lead to sinus problems and Parkinson's disease, but that does not make it in itself a disease. Not all or even most things that cause disease are diseases.

Perhaps the physicians are thinking that obesity shortens life, as actuarial tables indicate, and that is why it should be considered a disease. In addition to being vulnerable to the previous objection, this claim raises a more subtle problem. Even if obesity does shorten life, does it follow that it ought to be corrected? Physicians, as is well known, see their mission (their primary value) as preserving life. Nonphysicians, however, may value quality over quantity for life. Thus, even if I am informed – nay, guaranteed – that I will live 3.2 months longer if I drop forty-five pounds, it is perfectly reasonable for me to say that I would rather live 3.2 months less and continue to pig out. In other words, to define obesity as a disease is to presuppose a highly debatable valuational judgment.

Similar arguments can be advanced vis-à-vis alcoholism or gambling or child abuse. The fact that there may be (or are) physiological mechanisms in some people predisposing them to addiction does not in and of itself license the assertion that alcoholics (or gamblers) are sick. There are presumably physiological mechanisms underlying all human actions – flying off the handle, for example. Shall physicians then appropriate the management of temperament as their purview? (They have, in fact.) More to the point, shall we view people quick to anger as diseased – Doberman's syndrome?

Perhaps. Perhaps people would be happier if the categories of badness and weakness were replaced with medical categories. Physicians often argue that when alcoholism or gambling is viewed as sickness, that is, something that happens to you that you can't help, rather than as something wrong that you do, the alcoholic or gambler is more likely to seek help, knowing he or she will not be blamed. I, personally, am not ready to abandon moral categories for medical ones, as some psychiatrists have suggested. And, as Kant said, we must act as if we are free and responsible for our actions, whatever the ultimate metaphysical status of freedom and determinism. I do not believe that one is compelled to drink by one's physiological substratum, though one may be more tempted than another with a different substratum.

Be that as it may, the key point is that physicians are not discovering in nature that obesity or alcoholism are diseases, though they think they are. They are, in fact, promulgating values as facts and using their authority as experts in medicine to insulate their value judgments from social debate. This occurs because they do not see that facts and values blend here. They are not ill-intentioned, but they are muddled, as is society in general. And to rectify this, we must discuss, in a democratic fashion, which values will underlie what we count as health and disease, not simply accept value judgments from authorities who are not even cognizant of their existence, let alone conceptually prepared to defend them. At the very least, if we cannot engender a social consensus, we should articulate these for ourselves.

The Environmental Protection Agency has rejected scientifically sound toxicological data on moral grounds because the experiments were done by the Nazis on human beings against their will, out of fear of legitimating such experimentation.[7] This decision was made despite the fact that other well-established areas of science – such as research on hypothermia and human reactions to high altitude – are based foursquare on Nazi experiments and despite the fact that failing to use the data essentially entails that much invasive animal research will be done to replace it.

Indeed, perhaps most staggering to scientific ideology is the realization that the scientific revolution that ushered in modern science was rooted in revolutionary changes in values. Consider the development of modern physics at the hands of Galileo, Descartes, and Newton. While these thinkers did indeed lay the foundations for unprecedented changes in our empirical approaches to the world, a major part of their contribution involved pressing forward a new way of looking at reality, a new conceptual map of the furniture of the universe, or what one may rightly call a new metaphysic. While Aristotelian physics and its attendant view of the world stressed the need to explain the variety and irreducible qualitative differences found in experience – living

7 P. Shabecoff, "Head of the EPA Bars Nazi Data in Study on Gas," *New York Times,* March 23, 1988.

and nonliving, hot and cold, beautiful and ugly – a view still echoed in ordinary common sense, the new physics explained away these qualitative differences as by-products of changes in fundamental, homogeneous, mathematically describable matter, the only legitimate object of study. Modern physics correlatively *disvalued* the sensory, the subjective and the qualitative, and, as did Plato equated the measurable with the real. Thus, whereas Aristotle saw biology as the master science, and physics, as it were, the biology of nonliving matter, Descartes saw biology as a subset of physics and is therefore the herald of today's reductionistic molecular biology.

Or consider a revolution that I have looked at in considerable detail in another book: the replacement of psychology as the science of consciousness with behaviorism, which saw psychology as the science of overt behavior. What facts could force such a change – after all, we are all familiar with the existence of our subjective experiences. Few people were impressed with the denial of consciousness by behaviorism's founder, John B. Watson (he comes perilously close to saying, "We don't have thoughts, we only think we do"). Rather, people were moved by his valuational claims that studying behavior is more valuable, because we can learn to control it.

Clearly, then, the component of scientific ideology that affirms that science is value-free and ethics-free is incorrect. We can also see that the more fundamental claim – that science rests only on facts and includes only what is testable – is also badly wrong. How, for example, can one scientifically prove (i.e., empirically test) the claim that only the verifiable may be admitted into science? How can we reconcile the claim that science reveals truth about a public, objective, intersubjective world with the claim that access to that world is only through inherently private perceptions? How can we know that others perceive as we do or, indeed, perceive at all? (We can't even verify the claim that there are other subjects.) How can science postulate an event at the beginning of the universe (the Big Bang) that is by definition nonrepeatable, nontestable, and a singularity? How can we know scientifically that there is reality independent of perception? How can we know scientifically that the world wasn't created three seconds ago, complete with fossils and us with all our memories? How can we verify any judgments about history? How can we reply that we know things best when we reduce them to mathematical physics, rather than stay at the level of sensory qualities? And so on. Answers to the above questions are not verified scientifically. In fact, such answers are *presuppositional* to scientific activity.

Before closing our discussion, it is worthwhile to discuss briefly some other components of scientific ideology, which are so rarely discussed and so taken for granted by most scientists that I have called them collectively the "Common Sense of Science," for they are to science what ordinary common sense is to daily life.

We have in fact just alluded to another component of scientific ideology that worked synergistically with the denial of values to remove animal ethics from the purview of science. This is the claim that we cannot legitimately speak of thoughts, feelings, and other mental states in science, since we cannot deal with these things objectively, not having access to the thoughts and consciousness of others. As I have explained elsewhere, this denial allowed scientists to negate the reality of animal pain, distress, fear, and so on, while at the same time using animals as models to study pain! In a previous book,[8] I demonstrated that this viewpoint was adopted in the early twentieth century by behavioral psychologists despite the fact that the dominant approach to biology was Darwinian, and Darwin himself, and most of his followers, eloquently affirmed that if morphological and physiological traits were phylogenetically continuous, so too were mental ones. In fact, Darwin himself wrote a great deal regarding animal thought and feeling. I have shown in that book, I hope, that the removal of thought and feeling from legitimate science was not a matter of new data that refuted old attempts to study animal mind, nor was it a result of someone finding a conceptual flaw in that old approach (as Einstein did with Newton's views of Absolute space and time). In fact, the shift to studying behavior rather than mind was effected by valuational rhetoric, namely, that if we study behavior rather than thought, we can learn to shape it and modify it – to extract behavioral technology from science, as it were. Anyway, the rhetoric continued, real sciences such as physics deal with observables (a claim not always true), and if we want to be *real* scientists, we need to lose subjectivity. So despite the ideological belief that science changes only by empirical or logical falsification, we have shown that, at least

[8] Rollin, *The Unheeded Cry,* chapter 4.

in psychology, a major change in what counted as scientific legitimacy was driven by values.

Another component of scientific ideology that follows closely upon our discussion of values is the ubiquitous belief that we best understand any phenomenon when we have understood it at the level of physics and chemistry, ideally, physics. It is this component of scientific ideology that led a very prominent colleague of mine in physiology who works on fascinating issues in animal evolution at the phenotypic level to affirm in one of my classes that "science has passed me by. . . . My work is archaic. . . . All real science now operates at the molecular level."

This reductionistic approach further removes scientists from consideration of ethics. If what is "really real" and "really true" is what is described by physics, it is that much easier to treat ethical questions arising at the level of organisms as being as "unreal" or "untrue" as is the level at which they arise. The language of physics is, after all, mathematics; yet ethical questions seem inexpressible in mathematical terms. The belief that expressing things mathematically as physics does is getting closer to the truth leads in fact to a kind of "mathematics envy" among areas of science less quantitative, and sometimes leads to pseudo-mathematical obfuscation being deployed in fields such as sociology or psychology in order to make these fields appear closer to the reductionistic ideal. In the end, of course, as we pointed out regarding the scientific revolution, a commitment to reductionism represents a value judgment, not the discovery of new facts. No empirical facts force the rejection of the qualitative work for the quantitative, and Aristotle, for one, explicitly rejected such rejections.

The final element of scientific ideology worth mentioning is the belief that science should be ahistorical and aphilosophical. If the history of science is simply a matter of "truer" theories replacing false or partially false ones, after all, why study a history of superseded error? How things come to be accepted, rejected, or perpetuated is ultimately seen as not being a scientific question. Thus, many scientists lack a grasp of the way in which cultural factors, values, and even ethics shape the acceptance and rejection of whole fields of study (e.g., consciousness, as we have already discussed, eugenics, intelligence, race, psychiatry as a medical discipline, and so on). To take one very interesting example, it has been argued that quantum physics in its current form would never have been possible without the cultural context prevailing in Germany between 1918 and 1927.[9]

Historian Paul Forman has argued that a major impetus for both the development and the acceptance of quantum theory was a desire on the part of German physicists to adjust their values and their science in the light of the crescendo of indeterminism, existentialism, neoromanticism, irrationalism, and free-will affirmation that pervaded the intellectual life of Weimar and that was hostile to deterministic, rationalistic physics.[10] Thus quantum physicists were enabled to shake the powerful ideology of rationalistic, deterministic, positivistic late nineteenth- and early twentieth-century science, with its insistence on causality, order, and predictability, as a result of the powerful social and cultural ambience in German society that militated in favor of a world in which freedom, randomness, and disorder were operative and valued such chaos both epistemically and morally.

The rejection of philosophical self-examination is also built into scientific ideology and into scientific practice. Since philosophy is not an empirical discipline, it is excluded by definition by scientific ideology from the ken of science. Further, historically, philosophy, as did theology, competed with science, at least in the area of speculative metaphysics, so that the few historically minded scientists approach it with suspicion that is spread to others. In any case, scientists don't have time for "navel gazing" or "pontificating" as they often characterize philosophy – they are too busy doing science to reflect much on it. As one scientist said to me, "when I win a Nobel prize, then I will write philosophy, because then everyone will want to read it, whether it makes sense or not." Clearly then, reflection on science and ethics must also await a Nobel Prize.

[9] Ibid., p. 62.
[10] Forman, "Weimar Culture, Causality and Quantum Theory."

3

What Is Ethics?

Before we can explore the relationship between science and ethics, we must be clear about the general nature of ethics. This is particularly important in the age in which we live, since the rate of socio-ethical change has increased with great rapidity. As we shall see throughout our discussion, if professionals such as physicians, veterinarians, or researchers wish to keep their autonomy and steer their own ships, they must be closely attuned in an anticipatory way to changes and tendencies in social ethics and adjust their behavior to them, else they can be shackled by unnecessarily draconian restriction. And, as we saw in chapter 1 there have been numerous and bewildering ethical changes that professionals and others must adjust to throughout the second half of the twentieth century. There I catalogued some of the bewildering array of major socio-ethical changes that had developed in the second half of the twentieth century. As we shall see, failure of any subgroup in society to adjust to these ethical charges can result in major loss of freedom.

The first distinction that must be mastered is the difference between what I have called Ethics$_1$ and Ethics$_2$. Ethics$_1$, or morality, is the set of beliefs that society, individuals, or subgroups of society hold about good and bad, right and wrong, justice and injustice, fairness and unfairness. Ethics$_2$, on the other hand, is the logical examination, critique, and study of Ethics$_1$. What we are doing in this chapter and indeed in this book is Ethics$_2$. Though these are very different domains, it is easy to confuse them, both because we use the term "ethics" to apply to both and because sometimes moral philosophers and other moralists encroach on both domains. Martin Luther King, Jr., provides a good example of such an individual. Much of his work was a logical critique of segregationist moral principles as being conceptually incompatible with the basic principles laid out in the Declaration of Independence and the Constitution – freedom, equality, dignity of "all men." After all, if we "hold these truths to be self-evident, that all men are created equal," we cannot (as the Supreme Court pointed out in *Brown vs. the Board of Education*) treat members of different races differently, since different treatment entails unequal treatment. When making points such as this, King was functioning at the level of $Ethics_2$, though clearly he hoped thereby to change people's practices at the level of $Ethics_1$. On the other hand, when King preached nonviolence, he was actually pressing forward some new principles of ethics that he was hoping people would adopt, rather than pointing out that nonviolence was logically entailed by principles we already accept.

Similarly, what I am doing in this book is $Ethics_2$ – clarifying various ethical notions and showing their relevance to science – yet my purpose is clearly an attempt to get scientists to take $Ethics_1$ more seriously and to abandon the ideology we discussed that affirms that science is "ethics-free." But obviously, my ultimate goal is to make concern with $Ethics_1$ second nature to scientists and to change their behavior regarding moral issues.

Let us pause and look more carefully at $Ethics_1$ and $Ethics_2$. We have affirmed that $Ethics_1$ is the set of principles or beliefs that governs views of right and wrong, good and bad, fair and unfair, just and unjust. Whenever we assert that "killing is wrong," that "discrimination is unfair," or that "I think abortion is murder," one is explicitly or implicitly appealing to $Ethics_1$ – moral rules that one believes ought to bind society, oneself, and/or some subgroup of society such as physicians, veterinarians, researchers, or attorneys.

Under $Ethics_1$ therefore must fall a distinction among social ethics, personal ethics, and professional ethics. Of these, social ethics is the most basic and most objective, in a sense to be explained shortly.

People, especially scientists, are tempted sometimes to assert that unlike scientific judgments, which are "objective," ethical judgments are "subjective" opinion and not "fact," and thus not subject to rational discussion and adjudication. Although it is true that one cannot conduct experiments or gather data to decide what is right and wrong, ethics, nevertheless, cannot be based on personal whim and caprice. If anyone doubts this, let that person go out and rob a bank in front of witnesses, then argue before a court that, in his or her ethical opinion, bank robbery is morally acceptable if one needs the money.

In other words, the fact that ethical judgments are not validated by gathering data or doing experiments does not mean that they are simply a matter of individual subjective opinion. If one stops to think about it, one will quickly realize that very little ethics is left to one's opinion. Consensus rules about rightness and wrongness of actions that have an impact on others are in fact articulated in clear social principles, which are in turn encoded in laws and policies. All public regulations, from the zoning of pornographic bookstores out of school zones to laws against insider trading and murder, are examples of consensus ethical principles "writ large," in Plato's felicitous phrase, in public policy. This is not to say that, in every case, law and ethics are congruent – we can all think of examples of things that are legal yet generally considered immoral (tax dodges for the super-wealthy, for instance) and of things we consider perfectly moral that are illegal (parking one's car for longer than two hours in a two-hour zone).

But by and large, if we stop to think about it, there must be a pretty close fit between our morality and our social policy. When people attempt to legislate policy that most people do not consider morally acceptable, the law simply does not work. A classic example is, of course, Prohibition, which did not stop people from drinking, but rather funneled the drinking money away from legitimate business to bootleggers.

So there must be a goodly number of ethical judgments in society that are held to be universally binding and socially objective. Even though such judgments are not objective in the way that "water boils at 212°F" is objective (i.e., they are not validated by the way the world works), they are nonetheless objective as rules governing social behavior. We are all familiar with other instances of this kind of objectivity. For example, it is an objective rule of English that one cannot say, "You ain't gonna be there." Though people, of course, do say it, it is *objectively* wrong to do so. Similarly, the bishop in chess can objectively move only on diagonals of its own color. Someone may, of course, move the bishop a different way, but that move is objectively wrong, and one is not then "playing chess."

Those portions of ethical rules that we believe to be universally binding on all members of society, and socially objective, I call *social consensus ethics*. A moment's reflection reveals that without some such consensus ethic, we could not live together, we would have chaos and anarchy, and society would be impossible. This is true for any society at all that intends to persist: There must be rules governing everyone's behavior. Do the rules need to be the same for all societies? Obviously not; we all know that there are endless ethical variations across societies. Does there need to be at least a common core in all of these ethics? That is a rather profound question I address below. For the moment, however, we all need to agree that there exists an identifiable social consensus ethic in our society by which we are all bound.

Now, the social consensus ethic does not regulate all areas of life that have ethical relevance; certain areas of behavior are left to the discretion of the individual or, more accurately, to his or her *personal ethic*. Such matters as what one reads, what religion one practices or doesn't practice, how much charity one gives and to whom are all matters left in our society to one's personal beliefs about right and wrong and good and bad. This has not always been the case, of course – all of these examples were, during the Middle Ages, appropriated by a theologically based social consensus ethic. And this fact illustrates a very important point about the relationship between *social consensus ethics* and *personal ethics*. As a society evolves and changes over time, certain areas of conduct may move from the concern of the social consensus ethic to the concern of the personal ethic, and vice versa. An excellent example of a matter that has recently moved from the concern of the social ethic, and from the laws that mirror that ethic, to the purview of individual ethical choice is the area of sexual behavior. Whereas once laws constrained activities such as homosexual behavior, which does not hurt others, society now believes that such behavior is not a matter for social regulation but, rather, for personal choice, and thus social regulation of such activity withered away. About ten years ago the mass media reported, with much hilarity, that there was still a law on the books in Greeley, Colorado, a university town, making cohabitation a crime. Radio and TV reporters chortled as they remarked that, if the law were to be enforced, a goodly portion of the Greeley citizenry would have to be jailed.

On the other hand, we must note that many areas of behavior once left to one's personal ethic have been since appropriated by the social ethic. When I was growing up, paradigm cases of what society left to one's personal choice were represented by the renting or selling of one's real property and by whom one hired for jobs. The prevailing attitude was that these decisions were your own damn business. This, of course, is no longer the case. Federal law now governs renting and selling of property and hiring and firing.

Generally, as such examples illustrate, conduct becomes appropriated by the social consensus ethic when how it is dealt with by personal ethics is widely perceived to be unfair or unjust. The widespread failure to rent to, sell to, or hire minorities, which resulted from leaving these matters to individual ethics, evolved into a situation viewed by society as unjust and led to the passage of strong social ethical rules against such unfairness. As we shall see, the treatment of animals in society is also moving into the purview of the social consensus ethic, as society begins to question the injustice that results from leaving such matters to individual discretion. It is, of course, both possible and likely on some occasions that one's personal ethic may conflict with the social consensus ethic. I may personally believe that insults should be met by duels to the death, but I must still adhere to the social ethic's prohibition of my acting that way.

The third component of ethics, in addition to social consensus ethics and personal ethics, is *professional ethics*. Members of a profession are first and foremost members of society – citizens – and are thus bound by all aspects of the social consensus ethic not to steal, murder, break contracts, and so on. However, professionals – be they physicians, attorneys, researchers, or veterinarians – also perform specialized and vital functions in society. This kind of role requires special expertise and special training and involves special situations that ordinary people do not face. The professional functions that medical professionals perform also warrant special privileges – for example, dispensing medications and performing surgery. Democratic societies have been prepared to give professionals some leeway and assume that, given the technical nature of professions and the specialized knowledge their practitioners possess, professionals will understand the ethical issues they confront better than society does as a whole. Thus, society generally leaves it to such professionals to set up their own rules of conduct. In other words, the social ethic offers general rules, creating the stage on which professional life is played out, and the subclasses of society comprising professionals are asked to develop their own ethics to cover the special situations they deal with daily. In essence, society says to professionals, "Regulate yourselves the way we would regulate you if we understood enough about what you do to regulate you." Because of this situation, professional ethics occupies a position midway between social consensus ethics and personal ethics, because neither does it apply to all members of society, nor are its main components left strictly to individuals. It is, for example, a general rule of human medical ethics for psychiatrists not to have sex with their patients.

The failure of a profession to operate in accordance with professional ethics that reflect and are in harmony with the social consensus ethic can result in a significant loss of autonomy by the profession in question. One can argue, for example, that recent attempts to govern health care legislation is a result of the human medical community's failure to operate in full accord with the social consensus ethic. When hospitals turn away poor people or aged stroke victims, or when pediatric surgeons fail to use anesthesia on infants, or give less analgesia to adolescents than to adults with the same lesion,[1] they are not in accord with social ethics, and it is only a matter of time before society will appropriate regulation of such behavior. In veterinary medicine, as I discuss below, social fear of the irresponsible use and dispensing of pharmaceuticals threatened the privilege of veterinarians to prescribe drugs in an extralabel fashion (i.e., in a way not approved by the manufacturer) – a privilege whose suspension would have in a real sense hamstrung veterinarians. Because so few drugs are approved for animals, veterinary medicine relies heavily on extralabel drug use.

Below I demonstrate in detail how the scientific community has failed to accord with social consensus ethical expectations for the treatment of both animals and humans used as research subjects, largely as a result of that community having been blinded by the ideology I discussed that made ethics in science *invisible*. The result, of course,

[1] Walco, Cassidy, and Schechter, "Pain, Hurt and Harm."

was social dissatisfaction with science and correlative imposition of legislative constraints on scientific practice.

Let us return to $Ethics_2$: $Ethics_2$ is the logical, rational study and examination of $Ethics_1$, which may include the attempt to justify the principles of $Ethics_1$, the seeking out of inconsistencies in the principles of $Ethics_1$, the drawing out of $Ethics_1$ principles that have been hitherto ignored or unnoticed, engaging the question of whether all societies ought ultimately to have the same $Ethics_1$, and so on. This secondary sense of ethics – $Ethics_2$ – is thus a branch of philosophy. As we said, most of what we are doing in this book is $Ethics_2$, examining the logic of $Ethics_{1.}$ Socrates' activities in ancient Athens were a form of $Ethics_2$. Whereas we in society learn $Ethics_1$ from parents, teachers, churches, movies, books, peers, magazines, newspapers, and mass media, we rarely learn to engage in $Ethics_2$ in a disciplined, systematic way unless we take an ethics class in a philosophy curriculum. In one sense this is fine – vast numbers of people are diligent practitioners of $Ethics_1$ without ever engaging in $Ethics_2$. On the other hand, failure to engage in $Ethics_2$ – rational criticism of $Ethics_1$ – can lead to incoherence and inconsistencies in $Ethics_1$ that go unnoticed, unrecognized, and uncorrected. Although not everyone needs to engage in $Ethics_2$ on a regular basis, there is value in at least some people monitoring the logic of $Ethics_1$, be it social consensus ethics, personal ethics, or professional ethics. Such monitoring helps us to detect problems that have been ignored or have gone undetected and helps us to make ethical progress. Below I analyze some of the reasons why $Ethics_1$ is likely to stand in need of constant critical examination.

What is "philosophy" of which we have said $Ethics_2$ is a part? To tell someone that one is a "professional philosopher" or even a "philosophy teacher" is to risk a wide variety of undesirable responses, ranging from "Isn't everyone a philosopher?" to "Where is your couch?" to glassy-eyed stares to serious conversational lulls to questions about crystals and the prophecies of Nostradamus. To many of my nonphilosophy colleagues I am a sort of secular preacher, in whose company one refrains from telling off-color jokes, even though I do not refrain from telling them.

In fact, one can provide a fairly straightforward and clear account of philosophical activity that goes a long way toward breaking the stereotype and also helps people in any discipline understand why philosophy is relevant to them. As Aristotle long ago pointed out, all human activities and disciplines rest on certain assumptions and concepts that are taken for granted. As in the paradigmatic case of geometry, we must assume certain notions without proof, for it is on these notions that all subsequent proof is based. If we could prove our foundational assumptions, it would need to be on the basis of other assumptions that are either taken for granted or proved on the basis of other assumptions. Because the latter tack would lead to a never-ending hierarchy of assumptions and proofs – what philosophers call an infinite regress – certain things are simply assumed.

All disciplines and activities make such assumptions: Science assumes that we can identify causes and effects. Art assumes that certain objects are works of art and others are not. History assumes that we can reconstruct the past. Mathematics assumes that certain things count as proof and others do not. The law assumes that people are to be held responsible for certain actions wherein they acted freely. Some schools of ethology assume that animals are conscious beings; others assume that they are physiological machines.

In all of these examples it is obvious, on reflection, that certain basic challenges can be directed toward all of these implicit or explicit assumptions. What makes certain things works of art and others not? (Marcel Duchamp humorously asked this question when he submitted a porcelain urinal to a Paris sculpture exhibition in the early twentieth century.) Why do we accept someone's brain tumor as exonerating criminal behavior, but not someone's childhood experiences? What about that person's genes? How do we decide which of two incompatible but well-researched historical reconstructions to accept? How do we decide which approach to animal behavior is the correct approach? Is it the one that gives primacy to behavior, the one that gives primacy to evolutionary explanations, the one that invokes consciousness, or the one that invokes neurophysiology? Ought we or ought we not to accept a computer proof by exhaustion of Euler's conjecture? This is the terrain in which philosophy operates.

People who raise such basic, conceptual questions about fundamental concepts and assumptions are functioning as philosophers, whether or not they are professionally involved in philosophy. Because most people at some time or another ask such questions, most people have their philosophical moments. Much progress in human thought and behavior has been accomplished by such questioning of what is taken for granted by others. For example, one of Einstein's major contributions in developing special relativity was a philosophical critique of notions that physicists since Newton had taken for granted – namely, that one can talk intelligibly of absolute space, absolute time, absolute simultaneity, independently of who is recording or measuring these things. Indeed, it is for this reason that a major book on Einstein's work is entitled *Albert Einstein: Philosopher/Scientist.*[2]

As Plato noted,[3] what we assume about right and wrong, good and bad, justice and injustice, fairness and unfairness constitutes the most important assumptions we make as individuals, societies, or subgroups of societies, such as the professions. Our vision of the good, of what is right and wrong to do, underlies everything we do at all levels – be it the social level of policies about taxation and redistribution of wealth, which kind of science we do and don't fund (research into environmental preservation vs. research into the relationship between race and intelligence), our views of punishment and rehabilitation, and so on, or be it at the level of individual action.

Saying that $Ethics_1$ is subject to rational criticism is often met by skepticism, particularly from scientists. If we can't empirically disprove ethical claims, how can we criticize them? We have already seen one strategy for such criticism – that of Martin Luther King, Jr., who showed us that segregationist rules and laws were logically incompatible with our fundamental social/ethical commitments. If the social ethics could not in fact be rationally criticized, we could make no ethical progress.

The best way to criticize social ethics rationally – or indeed any ethics – was described in detail by Plato.[4] Plato explicitly stated that people who are attempting to deal with ethical matters rationally cannot *teach* rational adults, they can only *remind.* Whereas one can teach one's veterinary students about the various parasites of the dog and demand that they give the relevant answers on a quiz, one cannot do that with matters of $Ethics_1$, except insofar as one is testing their

[2] Schilpp, *Albert Einstein.*
[3] Plato, *Republic.*
[4] Plato, *Meno.*

knowledge of the social ethic as objectified in law – what they may not do with drugs, for example. (Children, of course, *are* taught $Ethics_1$.)

We have already given an example of reminding from Martin Luther King, Jr. In teaching this notion, I always appeal to a metaphor from the martial arts. One can, when talking about physical combat, distinguish between sumo and judo. Sumo, of course, involves two large men trying to push each other out of a circle. If a hundred-pound man is engaging a four-hundred-pound man in a sumo contest, the result is a forgone conclusion. In other words, if one is simply pitting force against force, the greater force must prevail. On the other hand, a hundred-pound man can fare quite well against a four-hundred-pound man if the former uses judo, that is, turns the opponent's force against him. For example, you can throw much larger opponents simply by "helping them along" in the direction of the attack on you.

When you are trying to change people's ethical views, you accomplish nothing by clashing your views against theirs – all you get is a counterthrust. It is far better to show that the conclusion you wish them to draw is implicit in what *they* already believe, albeit unnoticed. This is the sense in which Plato talked about "reminding."

As one who spends a good deal of my time attempting to explicate new social ethics for animals to people whose initial impulse is to reject it, I can attest to the futility of ethical sumo and the efficacy of moral judo. One excellent example leaps to mind. Some years ago I was asked to speak at the Colorado State University Rodeo Club about the new ethic in relation to rodeo. When I entered the room, I found some two dozen cowboys seated as far back as possible, cowboy hats over their eyes, booted feet up, arms folded defiantly, arrogantly smirking at me. With the quick-wittedness for which I am known, I immediately sized up the situation as a hostile one.

"Why am I here?" I began by asking. No response. I repeated the question. "Seriously, why am I here? You ought to know, you invited me."

One brave soul ventured, "You're here to tell us what is wrong with rodeo."

"Would you listen?" said I.

"Hell no!" they chorused.

"Well, in that case I would be stupid to try, and I'm not stupid."

A long silence followed. Finally someone suggested, "Are you here to help us think about rodeo?"

"Is that what you want?" I asked.

"Yes," they said.

"Okay," I replied, "I can do that."

For the next hour, without mentioning rodeo, I discussed many aspects of ethics: the nature of social morality and individual morality, the relationship between law and ethics, the need for an ethic for how we treat animals. I queried them as to their position on the latter question. After some dialogue they all agreed that, as a minimal ethical principle, one should not hurt animals for trivial reasons. "Okay," I said. "In the face of our discussion, take a fifteen-minute break, go out in the hall, talk among yourselves, and come back and tell me what *you guys* think is wrong with rodeo – if anything – from the point of view of your own animal ethics."

Fifteen minutes later they came back. All took seats in the front, not the back. One man, the president of the club, stood nervously in front of the room, hat in hand. "Well," I said, not knowing what to expect, nor what the change in attitude betokened. "What did you guys agree is wrong with rodeo?"

The president looked at me and quietly spoke: "Everything, Doc."

"Beg your pardon?" I said.

"Everything," he repeated. "When we started to think about it, we realized that what we do violates our own ethic about animals, namely, that you don't hurt an animal unless you must."

"Okay," I said, "I've done my job. I can go."

"Please don't go," he said. "We want to think this through. Rodeo means a lot to us. Will you help us think through how we can hold on to rodeo and yet not violate our ethic?"

To me that incident represents an archetypal example of successful ethical dialogue, using recollection and judo rather than sumo!

This example has been drawn from an instance that involved people's personal ethics; the social ethic (and the law that mirrors it) has essentially hitherto ignored rodeo. But it is crucial to understand that the logic governing this particular case is precisely the same logic that governs changes in the social ethic as well. Here also, as Plato was aware, lasting change occurs by drawing out unnoticed implications of universally accepted ethical assumptions.

The difference between judo and sumo (or reminding vs. teaching) in changing social ethics is well illustrated by comparing civil rights with Prohibition. Lyndon Johnson, in passing the civil rights acts, used the same logic we attributed to Dr. King. Himself a southerner, Johnson realized that most southerners would acquiesce to the premises "All humans should be treated equally" and "African Americans are humans." They just never put them together in a deductive argument. Johnson realized that if he "wrote this large" in law, people would realize their own commitments and acquiesce to the conclusion.

Had Johnson been wrong, civil rights would have been as irrelevant and ineffectual as Prohibition, where a small subgroup of society adroitly manipulated political power to ban drinking, surely a sumo move. Of course no one stopped drinking; in fact people drank *more,* because of the illicit thrill of patronizing speakeasies. All Prohibition did was create contempt for law, make ordinary citizens less respectful thereof, enrich the Kennedys and Canadian bootleggers, and give organized crime a foothold in legitimate business they never lost.

Personal and professional ethics are as open to rational criticism as are social ethics. Consider the following example: One's religious beliefs are certainly a matter of personal ethics, yet one can rationally criticize the contents of another's beliefs. For example, I often ask my audiences how many of them are Christians and, if they are, to hold up their right hands. I also ask the same audiences how many of them are ethical relativists, explaining that an ethical relativist is a person who believes that there are no objective ethical truths, that everyone's opinion is equally valid. I ask the relativists to hold up their left hands. Many people end up holding up both hands. But this is *logically* impossible! One can't be a Christian and a relativist at the same time, because a Christian must believe that certain things are absolutely right and wrong – for example, the Ten Commandments – whereas a relativist asserts that nothing is absolutely right or wrong.

In the same way, one can criticize a libertarian who also advocates censorship or a socialist who argues that the free market should control health care decisions. In fact, given the origins of people's personal ethics, chances are they will contain logical inconsistencies and incoherencies. One's opinions on sexual morality, for example, are shaped by parents, grandparents, schools, churches, peers, books, television, and movies. The minister may tell a teenager to abstain; the high school may advise the use of condoms; friends may counsel various forms of sexual release other than intercourse. President Clinton apparently did not see oral sex as real sex. And so on. So how likely is it that we will be consistent in our views? And this is true across all our ethics because we are typically not brought up to reflect philosophically on our ethics.

Professional ethics is certainly open to rational criticism. Two personal examples come to my mind. In the mid-1970s I wrote some articles criticizing veterinary ethics, as embodied in the American Veterinary Medical Association Code of Ethics, for failing to deal with many of the issues that society expected veterinary ethics to deal with, such as whether one should euthanize a healthy animal for owner convenience, when one should control pain in animals, whether veterinarians have a social obligation to lead in changing practices that hurt animals. Instead, much attention was devoted to questions of etiquette – how big one's sign can be, what form one's yellow pages ad should take, whether or not one can advertise, and so on.[5] As Dr. Harry Gorman (my co-teacher in veterinary ethics) pointed out to me, society got tired of the bickering about advertising, and the decision about its acceptability was made by the courts, not by veterinarians.

A second example concerns the treatment of animals in research. Also beginning in the 1970s, I attempted to persuade researchers who use animals that, though how they treated animals had essentially been left by society to the discretion of their professional and personal ethics, their behavior was not in accord with emerging social ethics on animal treatment and was, in fact, at odds with it. I argued that researchers were living on borrowed time. If society knew about some of the practices that were rife in research, such as the systematic failure to use analgesics for postsurgical pain, the multiple use of animals for invasive procedures in teaching, and general poor care, then society would appropriate the treatment of animals in science into the social ethic, no longer leaving it to the professionals. Sure enough, that is what occurred, and what needed to occur, and we discuss this in detail in a subsequent chapter.

In fact, as we mentioned at the start of this chapter, professionals should be zealous in seeking out – and listening to – rational criticisms of their ethics. Failure to do so can put them at loggerheads with social ethics, resulting in a loss of autonomy. An excellent example of this can be found in what happened some years ago with veterinary medicine.

[5] Rollin, "Updating Veterinary Ethics."

Veterinarians were at the time abusing their prescription privileges to provide a huge volume of antibiotics to farmers to use for growth promotion in farm animals. This activity was widely believed to be driving pathogen resistance to antibiotics and thereby endangering public health. Congress responded by floating legislation that would have removed the privilege of extralabel drug use from veterinarians. (Extralabel drug use means using drugs in a manner other than that which is indicated on the label.) Since most drugs used in veterinary medicine are in fact human drugs, removing extralabel drug use would have hamstrung veterinary medicine. This graphically illustrates how failure to accord with social ethics can severely curtail professional autonomy and power.

We have thus far explained how all ethics can be criticized by Ethics_2. As clear as the distinction between Ethics_1 and Ethics_2 seems to be, people get it wrong. This was humorously illustrated by the following incident.

I have been teaching veterinary ethics at the Colorado State University College of Veterinary Medicine for over twenty-five years. I've probably taught more than three thousand nascent veterinarians. And, like anything else, there are good years and bad years. One particularly bad year, I had spent a great deal of time explaining to the students that I was not there to teach them what was right and wrong – if they didn't know that by the time they reached vet school, they and society were in serious trouble. My job, I insisted, was to teach them how to *think about ethics,* how to recognize situations in veterinary medicine where subtle questions of right and wrong might arise and not be noticed; how to apply their notions of right and wrong to new situations; how to avoid contradicting themselves when doing so, and so on – in short, how to reason about right and wrong. This particular class wouldn't accept this. They wanted me, they said, to tell them exactly what was right and wrong. "We want answers," they whined. "You only give us questions!" Try as I might, I could not get them to see that my job was to help them think about such problems, not to dictate solutions.

One day, I had an inspiration. I came to class early and filled the blackboard with a series of statements: "Never euthanize a healthy animal." "Never crop ears or dock tails," and so on. When the students came into class, I pointed to the statements and told them to copy them down and memorize them. "What are they?" they queried. "These are the answers you've been asking for," I replied. "Who the hell are you to give us answers?" they shouted.

Now that the distinction between $Ethics_1$ and $Ethics_2$ has been made, we must briefly examine some major $Ethics_2$ issues directly related to science and ethics. Most prominent, of course, is the scientific ideological notion that, since ethical judgments cannot be verified empirically, they have no scientific meaning and are at best just individual opinion. Interestingly enough, Plato addressed this question well over two thousand years ago.[6] Plato pointed out that any group of individuals, even a group banded together for nefarious purposes such as a band of robbers, must presuppose some rules of interaction governing their conduct toward one another, or they could accomplish nothing. To take a simple example, such a group would surely have a moral rule against shooting one another if someone said something that made another angry. It is well known that the Mafia operates by strong moral principles of keeping confidences, being loyal, being obedient, and so on, as do even the most vicious youth gangs.

Thomas Hobbes pointed out that without some enforceable social order human life would be nasty, miserable, brutish, and short.[7] For Hobbes, we are all equal in being selfish and capable of killing and harming others. Even the weakest people can band together and kill the strongest at vulnerable times, such as when someone is asleep. Rationality therefore dictates that we surrender some of our freedom to a centrally enforceable set of rules, which I have called the social consensus ethic, in return for security for our life and property. In short, ethics is natural to the human situation, as a set of fences protecting us from each other.

Not everyone who theorizes about the basis for ethics is as negative and cynical as Hobbes. Plato, for example, saw ethics as not only protecting but fulfilling the potential in human life in a positive way.[8] No single person is good at everything relevant to fulfilling human desires. Some of us are craftsmen, others are "all thumbs"; some of us are musicians; others are tone deaf; some of us are strong; others weak. Imagine if we were all totally independent. Each of us would

[6] Plato, *Republic.*

[7] Hobbes, *Leviathan.*

[8] Plato, *Republic.*

be responsible for getting food and shelter, for example. To do so, we would each need to fish, hunt, grow crops, and build houses. We would in turn need tools to accomplish these purposes, which we would need to construct, and so on. It is easy to see that we are all far better off if we divide our labor. Instead of everyone doing everything, those good with their hands build, while those good at growing things raise food, while others fish or hunt, and so on. By division of labor in accord with our respective strengths, we all benefit. But to accomplish this, we need rules of conduct regulating our interactions. These rules making a good life possible are ethical edicts.

Plato points out that even at the level of the individual person, ethical rules are needed to balance the competing demands of different sides of human nature. My carnal appetites for food, sex, and intoxication vie against my desire for personal safety, which again competes with my desire for vengeance and my urge to protect my loved ones. If these competing urges are not ordered by consistent principles, life again becomes precarious or unfulfilled. If my gluttony dominates everything else, how will I assure my personal safety? If my anger dominates everything else, how will I be able to work with others helping me to fulfill my needs? The Icelandic epic poems vividly paint the black existence of those consumed by the desire for revenge; other works of literature such as Maugham's *Of Human Bondage* epitomize human downfalls through unbridled lust or pride or jealousy.

So it is fair to affirm that ethics is essential to human life, not only at any group level, but even at the level of the individual. Philosopher Thomas Reid pointed out[9] that even the most human of activities, language, rests foursquare upon an ethical presupposition of truth telling: "Veracity is a presupposition of discourse."

Another fundamental $Ethics_2$ question is the ancient problem, first raised by the Sophists, of ethical relativism. Ethical relativism, often affirmed by modern anthropologists, states that as one looks at different cultures one finds hugely varied opinions on what is moral and immoral. For example, as the Sophists pointed out, incest in Greece was a major moral transgression, whereas it was obligatory for the Egyptian royal family. Because different cultures vary widely in ethics, it is argued, we cannot give priority to any one ethical system. (Shadows

[9] Reid, *Inquiry into the Human Mind on the Principles of Common Sense.*

of such relativism can today be found in emphases on multiculturalism and diversity.)

In the first place, we must point out that the above version of the argument rests upon a logical error. The mere fact that there is a plurality or opinions on a given issue does not entail that there is no correct opinion. So the presence of multiple approaches to morality does not itself suffice to prove that there is no "true" ethic. The above argument must be revised to read as follows: (1) There is a plurality of ethical opinions. (2) There is no rational decision procedure (such as empirical confirmation) for adjudicating between them. Therefore, (3) We are not licensed to speak of ethical absolutes or of one ethical system being superior to another.

There exist various responses to ethical relativism.

(1) *Relativism is self-defeating.* Relativism asserts that all ethical positions are equally valid or true. In saying this, the relativist admits that his of her own position has no special validity and that the ethical position that *denies* the legitimacy of relativism is as true as relativism. Thus, if relativism is correct, its absolute correctness cannot be asserted by its defenders.

(2) *There exist criteria for judging competing ethics.* It is certainly true that there are differences in people's (and societies') ethical approaches. That in itself as we saw, however, does not mean that all approaches are equally valid. Perhaps we can judge different ethical views by comparing them to the basic purposes of ethics and to the reasons that there is a need for ethics in the first place.

As mentioned earlier in this discussion, rules for conduct are necessary if people are to live together – which of course they must. Without such rules, with people doing whatever they wish, chaos, anarchy, and what Hobbes called "the war of each against all" would ensue. Some idea of what such a world would be like may be gleaned from what happens during wars, floods, blackouts, and other natural or man-made disasters. A perennial source of fiction and drama, such situations lead to looting, pillaging, rape, robbery, outrageous black-market prices for such necessities as food, water, and medicine, and so on.

What sorts of rules best meet the needs dictated by social life? We know through ordinary experience and common sense what sorts of things matter to people. Security regarding life and property is one such need. The ability to trust what others tell us is another. Leaving certain things in one's life to one's own choices is a third. Clearly, certain moral constraints, principles, and even theories will flow from these needs. Rational self-interest dictates that if I do not respect your property, you will not feel any need to respect mine. Because I value my property and you value yours, and we cannot stand watch over it all the time, we "agree" not to steal and adopt this agreement as a moral principle. A similar argument could be mounted for prohibitions against killing, assault, and so on. By the same token, the prohibition against lying could be based naturally on the fact that communication is essential to human life and that a presupposition of communication is that, in general, the people with whom one is conversing are telling the truth, as we mentioned earlier.

By the same token, Kant emphasized, certain conclusions can be drawn about morality from the fact that it is based on reason. We would all agree that the strongest way someone can err rationally is to be self-contradictory. To be sensible, or rational, we must be consistent. According to some thinkers, something very like the Golden Rule is a natural consequence of a requirement for consistency. In other words, I can be harmed in a certain ways, helped in others, and wish to avoid harm and fulfill my needs and goals – and I see precisely the same features in you and the same concerns. Thus if I believe something should not be done to me, I am led by the similarities between us to conclude that neither should it be done to you, by me or by any other human being. In fact, it is precisely to circumvent this plausible sort of reasoning that we focus on differences between ourselves and others: color, place of origin, social station, heritage, genealogy, anything that might serve to differentiate you from me, us from them, so I don't have to apply the same concerns to others as to me and mine. The history of civilization, in a way, is a history of discarding differences that are not relevant to how one should be treated, such as sex or skin color. In sum, some notion of justice – treat equals equally – has been said to be a simple deduction from logic.

In support of this argument, one can say that at least a core of common principles survives even cross-cultural comparison. For example, some versions of the Golden Rule can be found in Judaism, Christianity, Islam, Brahmanism, Hinduism, Jainism, Sikhism, Buddhism,

Confucianism, Taoism, Shintoism, and Zoroastrianism. And it stands to reason that certain moral principles would evolve in all societies as a minimal requirement for living together. Any society with property would need prohibitions against stealing; communication necessitates prohibitions against lying; murder could certainly not be freely condoned; and so on.

Although I risk transgressing against current political correctness ideology, I would affirm that it does make perfect sense to say that certain cultures' moral systems are better than those of others. This seems intuitively obvious. Very few of us are prepared to say that cultures that perform clitoral mutilations on female children, practice infanticide, seize property arbitrarily, put people in concentration camps, permit rape, or discriminate on the basis of skin color are morally as good as cultures that eschew such practices. Intuitions, however, are not arguments. But I believe we can, using the insights of philosophers such as Plato, Hobbes, and Rawls, develop an argument justifying our intuitions.

Plato, at the end of the *Republic,* affirms that only a person who has grasped genuine, absolute morality can make a wise and rational decision about what sort of life to choose if one is reincarnated. One way of interpreting this Platonic myth or allegory is to say that when and only when one understands the role of ethical systems, one is in a position to know which systems are better than others. As we discussed earlier, one manifest purpose of ethical systems is to facilitate people living together effectively, since humans are social from birth. We also saw that, as Hobbes says, a primordial function of ethics is protecting people from each other, and another function, in a more positive vein, is facilitating cooperative efforts. If people willingly embrace a system of ethical rules, it must be in part because they are better off doing so than not doing so, and because they see those rules as a reasonable mechanism for *fairly* distributing the benefits and costs implicit in social life. Few of us would think an ethical system fair if it heavily favored a certain skin color, ancestry rather than accomplishment, or a particular religion.

Let us return to Plato's case: You are asked to choose in what ethical system you wish to be reincarnated. Additionally, as John Rawls has beautifully argued in his *Theory of Justice,* let us assume you do not know ahead of time your role in the system you choose. Whereas many of us would choose to live in pre-revolutionary France as an aristocrat or in ancient Greece as a citizen, few would choose those systems if we knew we would be slaves. If faced with such a choice, it would surely be most rational to choose a society where your fundamental interests were protected as much as possible regardless of your station in society, regardless of whether you were rich or poor, noble or commoner, white or black, male or female. If we look historically at the vast panoply of moral systems serving as the consensus ethic in different societies, we would have to conclude that current democratic societies represent the least arbitrary and fairest moral systems, and seem to be continuing to evolve in the direction of greater fairness. For this reason, I am quite comfortable emphatically rejecting ethical relativism. We shall shortly delineate the ethical theory underlying Western democratic societies.

First, however, we need to elucidate the process of making ethical judgments and decisions and recognizing ethical issues. Before an ethical issue can be resolved, it obviously must first be recognized as an ethical issue. This is not always easy. We perceive not only with our senses but with our beliefs, prejudices, theoretical commitments, expectations, values, perceptions, and acculturations.

When I teach this notion to students, I begin with the following child's trick:

I ask them to give me a single word for each thing I describe.
I say: What is a cola beverage that comes in a red can?
They say: Coke.
I say: If I tell a funny story, we call that a . . . ?
They say: Joke.
I say: If I puff on a cigarette, I . . .
They say: Smoke.
I say: I put some dirty clothes in a tub so they can . . .
They say: Soak.
I then say: What is the white of an egg called?
Most will automatically say: Yolk.

I go on to provide more serious examples of the ways that background, theory, and expectation can determine perception. The famous Rosenthal effect in psychology provides a nice scientific example.[10] Researchers studying rat behavior were told that one of the groups of

[10] Rosenthal, *Experimental Effects in Behavioral Research.*

white rats they would be working with was a special strain of highly intelligent rats. In subsequent studies, the researchers found that the bright rats did better than the ordinary rats in learning trials. In fact, they were all "ordinary" rats – the "brightness" came from the researchers' expectations. Often we experience the same "halo" effect with students in our classes, when we are told by other instructors of a particular student's brightness.

We can all recall the first time we looked at a radiograph. The radiologist pointed to what he said was a fracture, but we saw only dark and light, even though the same stimuli impinged upon our retinas as upon his. As one's knowledge of radiography broadens, however, one *sees* differently, though once again the retinal stimulation is unchanged.

Another amusing example is provided by a "paradox" that used to perplex people in the 1960s and 1970s called "The Boy with Two Fathers," which was presented as follows:

> A father and son are involved in an automobile accident. Both are seriously injured and are rushed to separate hospitals. The son is immediately readied for emergency surgery; at the first sight of him, however, the surgeon says, "I can't operate on this patient – he's my son!" How is this possible?

Twenty and thirty years ago one could perplex almost everyone in a class with this case. Today it falls flat – everyone sees the answer immediately: The surgeon is the boy's mother. Nothing in young people's expectations today precludes the possibility of a female surgeon.

Finally, let me cite a very poignant example from veterinary medicine. In the mid-1980s I was team teaching a veterinary ethics course with a prominent surgeon. I was discussing the tendency in veterinary medicine (and in science in general) through most of the twentieth century to ignore animal pain. In the midst of my lecture the surgeon stopped me. "My God," he said, "I was trained in the mid-sixties and was taught to castrate horses using succinylcholine chloride [a curariform drug that paralyzes but provides no anesthetic or analgesic effect]. This is the first time it ever dawned on me that the animals must have been hurting!" I return to the relationship between human mind-set and animal pain later in this book.

For now, however, the important point to realize is that the study of ethics provides a way of forcing people, on ethical matters, to go beyond their mind-set and expectations – indeed, that is why many people find it discomfiting. Of course, one can to some extent free oneself from the shackles of univocal perspective by seeking out people with strongly divergent opinions as discussion partners. But this goes counter to human nature; most people would rather deal with likeminded peers. A psychiatrist friend of mine working at a major Ivy League medical school did an extended study of interdisciplinary teaching. He found that genuine interdisciplinarity – team teaching for a whole semester with a colleague from a foreign discipline present at every class – is the most threatening academic experience for a faculty member, whereas superficial interdisciplinarity is easy, for example, if I come into your class for one lecture in a semester. Prolonged interdisplinarity, such as putting a philosopher and a clinician in one classroom for a semester, is very stress-inducing because the assumptions are so different. This tendency is, of course, reinforced by professional education. I recall a veterinarian colleague telling me that he was really only socially comfortable with fellow veterinarians who had graduated from a Midwestern veterinary school within about four years of his own graduation! Generally, medical students, veterinary students, and even graduate students are trained in a context militating in favor of "staying with their own kind."

A real ethical case I was involved with in my capacity as University Bioethicist at Colorado State University nicely illustrates the degree of difficulty people have with analyzing ethical issues in virtue of their training.

A man brought a small comatose dog with a head injury into our veterinary school clinic. He freely admitted, even boasted, that he had struck the dog in the head with a frying pan because it barked too much. When the dog did not regain consciousness, and the man's wife became upset, he took the dog to his regular practitioner. The veterinarian advised him to take the dog to the veterinary school hospital. The dog died there, and the animal's body was brought to necropsy and presented as a case to a group of students by a pathology instructor.

Coincidentally, one of the veterinary students in that class was an animal control officer, among whose duties was investigating cruelty complaints. With the instructor's permission, the student took the client's name from the file and began to investigate the case, phoning the client's home and speaking with his wife. The client became irate and complained to both the referring veterinarian and to the veterinary school clinician who had taken his case that his right to privacy had been violated. The private practitioner and the veterinary school referral clinician (the doctor to whom the case had been referred) in turn were furious with the student. The student was frightened, worried about the effect of the incident on his academic and subsequent career, and sought help.

What moral conflicts and problems does this case raise? Initially, the referring practitioner, the veterinary school clinician, and some administrators saw only one issue: the betrayal of client confidentiality by the student. (This is one of the few ethical issues that has been traditionally raised in veterinary education.) As the case evolved, administrators were also troubled by the involvement of the pathologist, who had "betrayed" the identity of the client. Only after much dialogue with me, the pathologist and the student did the parties begin to realize that there were many other concurrent issues.

First, there was an animal welfare issue: The client should not be allowed to fatally beat an animal with impunity. In addition, there was a social or moral obligation to report the occurrence of a crime, the same sort of moral obligation (now also a legal one in human medicine) that exists for health care professionals to report suspected child abuse. Furthermore, there was the moral (and legal) question of whether one could invoke confidentiality in a public teaching hospital, where it is implicit that cases will be discussed with students as part of their learning process. Last, the pathologist argued that, as a veterinary teaching institution, the school had a high moral obligation not to condone that which society as a whole has recognized as immoral and illegal.

Some veterinarians argued that the pathologist was within his rights to reveal the name but that the student ought not to have acted on the information. To this point the student replied that, as a law officer, he had a sworn duty (a moral obligation) to enforce the law. Some veterinarians hypothesized that if confidentiality isn't strictly observed, abusers of animals will not bring animals in for treatment. A controversy also arose over the fact that the school clinician had at least obliquely threatened the student with recriminations when he came to the clinic for his rotations. Others worried that the information about the case and these issues had not been sent back to the referring veterinarian for that party to handle. The issue of a conflict of interest between being a veterinary student and serving with animal control was also raised.

Ultimately, the situation was resolved, at least for future cases, by the university's drafting a formal policy that suspected abuse cases of this sort would automatically be reported to the hospital director and government authorities. One of the noteworthy features of the case was its dramatic teaching value in demonstrating just how complex a single ethical problem or case can be.

Given our earlier discussion of scientific ideology and our realization that we perceive in part through ideology, we can understand how scientists can fail to recognize ethical issues (or subjective experiences in animals) that are patent to ordinary common sense. Any twinges in the direction of ordinary common sense get suppressed; recall the story of the emperor's new clothes. If one's peer group says uniformly that animal use in research is not a moral issue but a scientific necessity, and one must accept this to receive the requisite education, such a belief becomes incorporated into the cognitive categories one uses to interpret the world.

As I just indicated, one way to counter this ideological acculturation is to solicit discussion partners from a wide variety of perspectives. While this sounds good, it is virtually impossible to effect. Which animal researcher will be comfortable enough to discuss his or her work with a member of People for the Ethical Treatment of Animals (PETA)? And if you are the rare person willing to do so, you will almost certainly be censured for "bringing our work to the activists' attention." In addition, the PETA person will not be comfortable with you, since from their perspective, you may well be an "animal torturer."

The failure to recognize other ethical perspectives (or *any* such perspectives in virtue of scientific ideology) is a major reason that the scientific community has been "blindsided" by social/ethical concerns it has failed to comprehend and will continue, as we shall see, to be so blindsided. One cannot respond to an ethical issue in a socially acceptable way until one at least sees it as an ethical issue.

Let us continue our discussion by assuming that one has analyzed an ethically charged situation and brought to light all of its relevant components. How does one then proceed?

There are, of course, many situations that may have diverse ethical components yet whose resolution is clearly dictated by the social ethic.

If a person owes me money and I spot him on the street, I cannot jump him, beat him senseless, and search his wallet for my just compensation. Similarly, if I believe that a person is not raising his or her child properly, I cannot kidnap the child. Even when the social ethic does clearly dictate positions, scientific or other ideology may eclipse them. As we see below in our discussion of human research, certain medical experiments done on humans in the United States were clearly forbidden by the social ethic, at least by its spirit, yet were performed anyway.

Some situations do not at all involve the social ethic, and must instead be resolved by one's personal ethic. For example, whether or not to use former pets in research where using them is not forbidden by law becomes a decision for the researcher's own personal ethic. Other situations require a profession as a whole to take a stand, as when U.S. science journal editors decided not to publish results detailing animal research done in other countries in ways that would have been rejected by U.S. law and policy, even though it was legal to perform the research in the other country. (I myself sometimes perform such "ethical review" for journals whose editors suspect that a foreign study would not have been permitted here.)

How does one rationally make an ethical decision not dictated by the social ethic, particularly a personal ethical decision? There are some fundamental notions to recall. Let us take a hint from the philosopher Wittgenstein and ask ourselves how we learned the words "right" and "wrong," "good" and "bad," and other moral elements of our conceptual toolbox. When we were small, we might have finished our dessert (chocolate pudding, my personal favorite) and reached over to take our little brother's dessert. We are interrupted when mother slaps our hand and declares in no uncertain terms "No! That is wrong" (or "bad"). We may perhaps initially believe that this prohibition applies only to chocolate pudding, so we try again with ice cream, and are again rebuked. Eventually, we grasp the more abstract generalization that taking something from someone else without permission is wrong, to the even more abstract notion that stealing is wrong. In other words, we ascend from particulars to generalizations in our moral beliefs, just as we do in our knowledge of the world, moving from "Don't touch this radiator" to "Don't touch any hot objects" to "Hot objects cause burns if touched."

Let us call the ethical generalizations that we learn as we grow *moral principles* (or $Ethics_1$ principles). Although we originally learn such moral principles primarily from parents, as we grow older, we acquire them from many and varied sources – friends and other peers, teachers, churches, movies, books, radio and television, newspapers, magazines, and so on. We learn such diverse principles of "It is wrong to lie," "It is wrong to steal," "It is wrong to hurt people's feelings," "It is wrong to use drugs," "Stand up for yourself," and, of course, many others. Eventually, we have the mental equivalent of a hall closet chock-full of moral principles, which we (ideally) pull out in the appropriate circumstances. So far this sounds simple enough.The trouble is that sometimes two or more principles fit a situation yet patently contradict one another. It is easy to envision a multitude of situations wherein this dilemma might occur.

For example, we have all learned the principles not to lie and not to hurt others' feelings. Yet these may stand in conflict, as when a co-worker or wife asks me, "What do you thank of my new three-hundred-dollar hairdo?" and I think it is an aesthetic travesty. Similarly, many of my male students growing up on ranches also face such tension. On the one hand, they have been brought up as Christians and taught the principle "Turn the other cheek." On the other hand, they have also been taught "not to take any crap and to stand up for yourself." As a third example, a female colleague tells of suffering a great deal of anguish when dating, as she had been taught both to be chaste and not to make others feel bad.

When faced with such conflicts, many of us simply don't notice them. As one of my cowboy students said to me once about the internal conflict between turning the other cheek and not being bullied: "What's the problem, doc? 'Turn the other cheek' comes out in church, the other one comes out in bars." Obviously, this response is less than satisfactory.

The key to resolving such contradictions lies in how one prioritizes the principles in conflict. Obviously, if they are given equal priority, one is at an impasse. So one needs a higher-order theory to decide which principles are given greater weight in which sorts of situations and to keep us consistent in our evaluations, so that we do the same sorts of prioritizing in situations that are analogous in a morally relevant way. In this regard one can perhaps draw a reasonable analogy between levels of understanding in science (i.e., knowledge of the world) and ethics. In science one begins with individual experiences (e.g., of a moving body); one then learns a variety of laws of motion (celestial motion, Kepler's laws, terrestrial motion), and one finally unifies the variegated laws under one more general theory from which they can all be derived (Newton's theory of universal gravitation). Similarly, in ethics one begins with awareness that particular things are wrong (or right), moves to principles, and then ascends to a theory that prioritizes, explains, or provides a rationale for both having and applying the principles. Theories can also help us to identify ethical components of situations wherein we intuitively surmise there are problems but can't sort them out.

Ethical theorizing, the construction of ethical theories, is as old as human civilization. Probably the simplest form of ethical theory is to say that certain rules are valid because God said so – witness the Jewish canon, a plethora of principles. Such a view, of course, is of little help, since conflicts arise between the principles, and alluding to God's authorship doesn't help. In fact, one need only read the Talmud to find out how much endless controversy even allegedly unequivocal principles generate.

Construction of ethical theories has occupied philosophers from Plato to the present. It is beyond the scope of this discussion to survey the many diverse theories that have been promulgated. On the other hand, it is valuable to look at some significantly different systems that nicely represent extremes in ethical theory and that, more important, have been synthesized in the theory underlying our own consensus social ethic.

Ethical theories tend to fall into two major groups: those stressing goodness and badness, that is, the results of actions, and those stressing rightness and wrongness, or duty, that is, the intrinsic properties of actions. The former are called *consequentialist*, or *teleological*, theories (from the Greek word *telos*, meaning "result," "end," or "purpose"). The latter are termed *deontological* theories (from the Greek word *deontos*, meaning "necessity" or "obligation") – in other words, what one is obliged to do. The most common deontological theories are theologically based, wherein action is obligatory because commanded by God.

The most well-known consequentialist theory is *utilitarianism*. It has appeared in a variety of forms throughout history but is most famously associated with nineteenth-century philosophers Jeremy Bentham and John Stuart Mill.[11] In its simplest version utilitarianism holds that one acts in given situations according to what produces the greatest happiness for the greatest number, wherein *happiness* is defined in terms of pleasure and absence of pain. Principles of utilitarianism would be generalizations about courses of action that tend to produce more happiness than unhappiness. In situations wherein principles conflict, one decides by calculating which course of action is likeliest to produce the greatest happiness. Thus, in the trivial case of the ugly hairstyle mentioned previously, telling a "little white lie" will likely produce no harm, whereas telling the truth will result in hostility and bad feeling, so one ought to choose the former course of action.

Despite the intuitive appeal of utilitarianism, it raises many problems. Among these problems are how to measure pleasure and pain, how to compare pleasure and pain qualitatively, among different kinds of pleasure and pain. For example, consider sexual versus intellectual pleasure. If one refers to "higher pleasures," as Mill does, something besides *amount* of pleasure is ultimately good, or alternatively, if one affirms that the pleasures of sex and of reading differ only in degree, this seems wrong not only intuitively, but also because such a view can lead to *Brave New World* scenarios where everyone is kept happy with drugs or electro-stimulation of brain pleasure centers. Finally, "counting heads" for the benefit of the majority leads to the problem of oppression of the minority and totalitarianism. If I can make many people happy by seizing the property of a minority (as the Nazis did), why not do so?

There are of course other consequential theories – Plato's, for example, in the *Republic,* where he argues that the ultimate principle for judging good actions is what is conducive to social order and efficiency. But utilitarianism, in part for its intuitive appeal, has been most influential. The major point to bear in mind about utilitarianism is that, like all ethical theories, it provides a way of resolving conflict among principles and providing a higher rule for decision making.

Those among us who grew up with very liberal parents will quickly recognize the utilitarian approach. Suppose you approach such parents in a quandary. You are thinking of entering into an adulterous

[11] Bentham, *An Introduction to the Principles of Morals and Legislation*; Mill, *Utilitarianism.*

relationship with a married woman. You explain that she is terminally ill – despised and abandoned by her vile, abusive husband who does not care what she does, but who nonetheless sadistically blocks a divorce – and she is attempting to snatch a brief period of happiness before her demise. These parents might well say, "Adultery is generally wrong, as it usually results in great unhappiness. But in this case perhaps you both deserve the joy you can have together.... No one will be hurt."

On the other hand, those among us who grew up with German Lutheran grandparents can imagine a very different scenario if one approached them with the same story. They would be very likely to say, "I don't care what the results will be – adultery is always wrong! Period!" This is, of course, a strongly deontological position.

The most famous rational reconstruction of such a position is to be found historically in the writings of the German philosopher Immanuel Kant.[12] According to Kant, ethics is unique to rational beings. Rational beings, unlike other beings, are capable of formulating universal truths of mathematics, science, and so on. Animals, lacking language, simply do not have the mechanism to think in terms such as "all X is Y." As rational beings, humans are bound to strive for rationality in all areas of life. Rationality in the area of conduct is to be found in subjecting the principle of action you are considering to the test of universality, by thinking through what the world would be like if everyone behaved the way you are considering behaving. Kant called this requirement "the categorical imperative," that is, the requirements of all rational beings to judge their intended actions by the test of universality. In other words, suppose you are trying to decide on whether you should tell a little white lie in an apparently innocuous case, like the ugly hairdo dilemma. Before doing so, you must test that action by the categorical imperative, which enjoins you to "act in such a way that your action could be conceived to be a universal law." So before you lie, you conceive of what would occur if everyone were allowed to lie whenever it was convenient to do so. In such a world the notion of telling the truth would cease to have meaning and thus so, too, would the notion of telling a lie. In other words, no one would trust anyone.

[12] Kant, *Foundations of the Metaphysics of Morals.*

Thus, universalizing telling a lie leads to a situation that destroys the possibility of the very act you are contemplating, and therefore becomes rationally indefensible, *regardless of the good or bad consequences in the given case.* By the same token, subjecting your act of adultery to the same test shows that if one universalizes adultery, one destroys the institution of marriage, and would thereby in turn render adultery logically impossible. Thus, in a situation of conflicting principles, one rejects the choice that could not possibly be universalized.

Kant proceeds to draw some interesting corollaries from his analysis. In particular, in a manner I have discussed elsewhere,[13] he claims to show that it logically follows from the categorical imperative that one must treat all human beings as "ends in themselves," not merely as means, as possessing "intrinsic value," not merely "instrumental value." Let us attempt to explicate these obscure concepts in a manner useful to our discussion and grounded in common sense.

Let us imagine having clogged drain pipes. I may try to unclog them with a plunger from my garage. Suppose the plunger does the job. If that is the case, the plunger is a means to my end – unclogging pipes. I then am not morally obliged to reward the plunger; it is a tool to further my (or some other person's) ends or goals. Not only am I not obliged to reward it, I may throw it away, use it to start a fire, or toss it back into the garage without a "thank you." It is solely a means to my ends, possessing use value or instrumental value but no *intrinsic* value; since it is nonconscious, inert matter, what I do to it *doesn't matter to it*; it is incapable of valuing what happens to it, whether good or bad.

On the other hand, suppose, after trying the plunger, I need to call a plumber. Although the plumber is indeed a means to my end (unclogging my pipes), he *is not merely* that – he is a locus of moral attention since what I do to him matters to him. So it would be wrong for me not to pay him what I agreed to pay or to toss him into the garage or trash heap. I am obliged to behave morally toward him. He is not merely a tool – a means to achieve my ends – but he has needs, interests, desires, intentions, and feelings, the fulfillment or thwarting of which matter to him. Built into him is the positive and negative valuing of what is done to him; such valuing is intrinsic (built into him), rather than merely a result of how well he serves me or

[13] Rollin, "There Is Only One Categorical Imperative."

how much or how little I value his usefulness to me (his instrumental value). Thus, he is an "end in himself."

These Kantian concepts probably apply most naturally to our relationships of love, sex, and friendship with other people. We all recall how bad we can feel when we find out that someone we consider a friend maintained a relationship with us only because we had a car. We feel they used us as a means to their ends, without really caring about our ends.

Similarly, when sexual morality began to be thought about in secular terms in the 1960s, and the question of which sort of sexual relations were moral or immoral was raised outside a theological context, many people argued that what makes sex moral or immoral was not any given act, but rather how one views and treats one's partner. If my partner is merely an outlet for my lust, if I do not consider her pleasure or satisfaction, then I behave immorally toward her, for I see her solely as a *means,* possessing *instrumental* value. Thus even basic sexual intercourse in the "missionary position" can be immoral, whereas what has been called "perverse" can be perfectly moral if I respect my partner's needs and desires.

Like utilitarianism, Kantianism can be criticized in many ways. For example, it follows from his theory that since one ought never lie, one could not lie to spare someone's feelings on their deathbed, or even to prevent a maniac from initiating a bloodbath say if he asks, "Didn't I leave my Uzi with you?" Far from according with our moral intuitions, then, his theory flagrantly violates them. And if utilitarianism can lead to totalitarianism, Kantianism can lead to excessive moral emphasis on individuals with little regard for social benefit or needs.

There are many other ethical theories, including those called *virtue ethics,* emphasizing, as the Greeks did, the development of character traits in citizens. These provide valuable insights into many aspects of ethics. But for our purposes, it suffices to focus on Kantianism and utilitarianism. Both, it seems to me, are based in powerful and legitimate insights about features of moral life. Utilitarianism stresses the point that a key feature of ethics is its cognizance of the general welfare in society, the good of the group or the majority or the society as an entity (in this sense, Plato has been called a utilitarian). Kantianism, in contradistinction, focuses on the individual and his or her freedom and autonomy and reason as the direct object of moral attention. The reason that both theories do not fare well as general theories of ethics is that each takes one aspect of moral life and gives primacy to it at the expense of other equally important aspects. As we said, utilitarianism does not devote enough attention to protection of the individual and can indeed submerge the individual and rapidly become totalitarian. It also tends to emphasize feeling (pleasure) at the expense of reasoned autonomy or freedom, hence the *Brave New World* problem we raised. On the other hand, Kantianism does not pay enough heed to what is today called communitarian concerns, given Kant's concern with individual extreme autonomy. Hence the absurdities of trying to imagine a genuinely Kantian practicable social ethic. Kant also fails to realize that in real moral life, consistency and rule-governed behavior need to be tempered by mercy. Probably neither theory pays enough attention to many other aspects of moral life, such as special obligations we have to our families and others. Though these theories can be made compatible with such points by creating cumbersome ethical machinery, they do not do so comfortably. Neither theory, in my view, devotes enough attention to moral psychology, a major eighteenth- and early nineteenth-century concern.

So I do not believe, a priori, that either theory makes for an ideal practicable social ethic, though either can guide us in our individual personal ethical decision making. It is therefore no surprise that our American social consensus ethic as embodied in our constitutional democracy, and indeed the social ethics of all Western democratic societies, which I have argued earlier represent the best sorts of social ethics, do not fall at either end of the deontological/utilitarian spectrum. Not surprisingly, given that social ethics needs to work in the real world, not just to be conceptually consistent and coherent, the ethical theory adopted in democracies is a *mixture* of consequentialist/utilitarian notions and Kantian/deontological notions.

Whatever theory we adhere to as individuals, we must be careful to assure that it fits the requirements demanded of morality in general: It must treat people who are relevantly equal equally; it must treat relevantly similar cases the same way; it must avoid favoring some individuals for morally irrelevant reasons (such as hair color); it must be fair and not subject to whimsical change. In the same vein, obviously, a society needs some higher-order theory underlying its social consensus ethic. Indeed, such a need is immediately obvious in almost all social decision making, be it the military demanding life-threatening service from citizens or the legislature redistributing wealth through taxation. It is in society's interests to send you to war – it may not be in yours, as you risk being killed or maimed. It is in society's interest to take money from the wealthy to support social programs or, more simply, to improve quality of life for the impoverished, but it arguably doesn't do the wealthy individual much good.

Different societies have of course constructed different theories to resolve this conflict. Totalitarian societies have taken the position that the group, or state, or Reich, or however they formulate the corporate entity must unequivocally and always take precedence over the individual. The behavior of the Soviet Union under Stalin, Germany under Hitler, China under Mao, and Japan under the emperors all bespeak the primacy of the social body over individuals. On the other end of the spectrum are anarchistic communes, such as those of the 1960s, that give total primacy to individual wills and see the social body as nothing more than an amalgam of individuals. Obviously, societies along the spectrum are driven by different higher-order theories.

In my view, our Western democratic societies have developed the best mechanism in human history for maximizing both the interests of the social body and the interests of the individual. Although we make most of our social decisions by considering what will produce the greatest benefit for the greatest number, a utilitarian/teleological/consequentialist ethical approach, we skillfully avoid the "tyranny of the majority" or the submersion of the individual under the weight of the general good. We do this by considering the individual as, in some sense, inviolable. Specifically, we consider those traits of an individual that we believe are constitutive of his or her *human nature* to be worth protecting at almost all costs. We believe that individual humans are by nature thinking, speaking, social beings who do not wish to be tortured, who want to believe as they see fit, who desire to speak their mind freely, who act freely and make choices, who have a need to congregate with others of their choice, who seek to retain their property, and so forth. We take the human interests flowing from this view of human nature as embodied in individuals and build protective, legal/moral fences around them that insulate those interests even from the powerful coercive effect of the general welfare. These protective fences guarding individual fundamental human interests even against the social interest are called *rights.* Not only do we as a society respect individual rights, we do our best to sanction other societies that ride roughshod over individual rights.

In essence, then, the theory behind our social ethic represents a middle ground or synthesis between utilitarian and deontological theories. On the one hand, social decisions are made and conflicts resolved by appeal to the greatest good for the greatest number. But in cases wherein maximizing the general welfare could oppress the basic interests constituting the humanness of individuals, general welfare is checked by a deontological theoretical component, namely, respect for the individual human's nature and the interests flowing therefrom, which are in turn guaranteed by rights, a thoroughly Kantian idea.

The practical implications of this theory are manifest. Consider some examples. Suppose a terrorist has planted a bomb in an elementary school, placing the lives of innocent children in jeopardy. Suppose further that there is no way to defuse the bomb without setting it off unless the terrorist, whom we have in custody, tells us how to do so. But he refuses to speak. Most of us would advocate torturing the terrorist to find out how to neutralize the bomb; after all, many innocent lives are at stake. Yet despite the enormous utilitarian costs, our social ethic would not allow it, because the right not to be tortured is so fundamental to human nature that we protect that right at whatever cost.

Similarly, suppose I wish to give a speech advocating homosexual, atheistic, satanic bestiality in a small ranching community in Wyoming. The citizens do not wish me to speak – they fear heart attacks, enormous expenses for police protection, harm to children exposed to these ideas, and other evils. No one in the community wishes to hear me. Despite all this, I could call the ACLU or some such organization, and eventually federal marshals would be dispatched at enormous taxpayer expense to assure my being permitted to speak, even if no one in fact attended my speech.

This, then, is a sketch of our underlying social ethical theory. One may choose any personal ethical theory, but it must not conflict with the precedence of the social ethical theory. Thus I may choose to limit what I read by virtue of my adherence to some theological ethical theory, but if I am a librarian, I cannot restrict what *you* read.

There are of course many other aspects of ethics worthy of discussion. Our account has been designed to place the issue of science and ethics in a clear context. It is within this context that we discuss the particular tensions and problems that scientific ideology has engendered regarding scientific activity and social ethics.

4

Ethics and Research on Human Beings

If one imagines an extraterrestrial researcher looking at the history of science and ethics in the twentieth century, one can imagine such an observer not being surprised at, and even having some sympathy for, the failure of the scientific community to engage issues of animal research and toxicology testing, animal cloning, and genetic engineering of animals. After all, he or she might affirm, there has been relatively little thought devoted to ethics and animals – most of what there is has been a product of the last quarter of the twentieth century, and it takes time for such new ideas to be incorporated into social thought.

On the other hand, such a detached observer would almost certainly be shocked at the cavalier use of human beings in research during the same era, a use that in many cases was violative of absolutely fundamental social ethical commitments that have been thoroughly discussed for hundreds and even thousands of years. In an extraordinarily clear and perceptive statement made at the close of the trial of the infamous Nazi physicians who cavalierly used large number of prisoners, slave laborers, and concentration camp inmates in painful experiments, the chief prosecutor affirmed that "the most fundamental tenet of medical ethics and human decency [requires that] the subjects volunteer for the experiment after being informed of its nature and hazards."[1]

[1] Quoted in Moreno, *Undue Risk*, p. 79.

Notice that in addition to citing field-specific "medical ethics" the prosecutor also refers to *fundamental human decency*. Even among the most unsophisticated of citizens, one would find a queasiness at forcibly hurting some people – including particularly children, who are paradigmatically innocent – for the benefit of others. In other words, one need not be a scholar of medical ethics to find abhorrent Nazi experiments that, without consent or benefit to the subjects, forced forty to sixty gypsies to be starved and given only seawater to drink until they tried to drink dirty water used to mop floors, that involved forcibly inflicting gunshot and other wounds on concentration camp inmates to study healing, or that killed children to harvest their organs. These (and thousands of other Nazi experiments) are so violative of so many Western ethical principles that even a young child would readily perceive them as "wrong." How then, is it possible that thousands of well-educated people in one of Europe's most civilized countries could perform them without a qualm?

There are no simple magic formula answers to this query. After all, Kantian thinking – "treat people as ends in themselves" – permeates our ethical thought, as do fairness, justice, freedom, not harming innocents, and so on. As Robert Jay Lifton[2] and Daniel Goldhagen[3] have shown, pervasive ideology can override common sense and common decency. We have seen this occur countless times in human history, as when the Inquisition tortured and killed hundreds of thousands "for their own good" or when people were killed to save their immortal souls from damnation.

Medical scientists and physicians were among the earliest and most loyal supporters of Nazism. Nazism in its ideology audaciously equated the health of the body of the state and society with the health of individual organisms and gave a prominent role to medical scientists and physicians in assuring that health and in purging "pathogens." Jews, gypsies, genetically or mentally defective people, and others were seen as pathogens in the body politic, sapping its health and resources just as germs do to an individual. Though put into action by Hitler, the ideology of "racial hygiene," as it came to be called, had a long history in Germany, with certain races seen as harmful and inferior and,

[2] Lifton, *Nazi Doctors*.

[3] Goldhagen, *Hitler's Willing Executioners*.

above all, parasitic upon the healthy society and in need of purging and killing. Just as a physician must overcome natural human aversion to removing fecal impactions and handling pus, blood, and vomit, so too Nazi physicians and researchers overcame their natural reluctance to kill or hurt innocents, particularly women and children.

Lifton refers to coming to see killing (or harming) apparent innocents as a "therapeutic imperative."[4] He cites a Nazi doctor's response to the query, "How can you reconcile [killing] with your Hippocratic oath as a doctor?" "Of course I am a doctor and I want to preserve life," he replied. "Out of respect for human life, I would remove a gangrenous appendix from a diseased body. The Jew is the gangrenous appendix in the body of mankind."[5]

With unconscionable oversimplification we may say that Nazi ideology amplified and resonated with themes already prevalent in German culture. In the same way that ordinary physicians learn to overcome compunctions against hurting in order to heal, as when one does surgery, lances a boil, or removes a tumor-ridden organ, nonpsychopathic Nazi physicians overrode their fundamental decency with the ideology of racial hygiene and with beliefs that stressed that sometimes one needed to kill individuals to effect corporate health, and thus that some life was "unworthy of life."[6] Some notion of the exonerative power of such ideology may be gleaned from the fact that even the nonmedical killers of the *Einsatzkommando,* the extermination groups who exterminated Jews and others in Eastern Europe, generally thought to be sadists and psychopaths, suffered sleeplessness, madness, alcoholism, and even suicide after shooting women and children, so much so that Daniel Goldhagen reports that, for reasons of morale, eventually no one was *forced* to do such work.[7]

Let us stress again that however bizarre, unbelievable, and implausible such Nazi ideology may be to us, we have ample evidence that it in fact was readily accepted in Germany. Ideology is, of course, not the whole explanation; jealousy of the success of Jews in medicine, for example, clearly was a factor that helped to spur hatred among

[4] Lifton, *Nazi Doctors*, p. 15.
[5] Ibid., p. 16.
[6] Ibid., p. 27.
[7] Goldhagen, *Hitler's Willing Executioners.*

physicians and scientists. But since all of us in our professional or political or religious lives must buy into a set of assumptions, and these assumptions in part determine how we perceive, we can begin to realize how cultural and historical factors can make bright people like ourselves be held in thrall by what we – or others – may initially see as absurd.

When we are in the thrall of ideology, we don't realize it. Other ideologies look silly to us; our own is not even evident. Conversations with people who have experienced conversions from religious ideology to atheism or vice versa have led me to understand that, after the conversion, people cannot even recall what their preconversion state was *like*. Thus it does no good to ask, "How could those Germans believe that kind of garbage?"

There is also a widespread belief among ordinary U.S. citizens that Nazi science was demented in its aim and of no scientific value. Tabloids have given examples of the attempts to "Aryanize" non-Aryans by changing eye color or to waken people from cold-induced torpor by copulation. In fact, much Nazi science is exactly what one would expect from scientists with an unlimited pool of human subjects and an ideological barrier to feeling compassion or moral concern for them.

Far from being useless and demented, some of the most horrible Nazi research has been some of the most scientifically valuable. Hypothermia experiments orchestrated by Sigmond Rascher in Dachau have served as the experimental data presuppositional to later work.[8] By the same token, the high altitude experiments wantonly performed on prisoners served as the basis for the U.S. military's establishing the School of Aviation Medicine in San Antonio, Texas, with Nazi scientists such as Hubertus Strughold brought to the United States under a program called Operation Paper Clip, which also recruited rocket scientist Wernher Von Braun, who had been responsible for the death of at least 20,000 slave laborers.[9] Scientifically valuable toxicological data was also obtained from poisoning prisoners, which later raised the question of whether it was moral to use such data; the Environmental Protection Agency refused to use it.[10]

[8] Hunt, *Secret Agenda.*
[9] Ibid.
[10] Shabecoff, "Head of EPA Bars Nazi Data."

Whether the Nazi work was scientifically valuable or not, the Nuremburg trial of the Nazi researchers awakened worldwide revulsion. In issuing their judgment against the scientists involved, the court issued what has come to be called the Nuremberg Code, a codification of the minimal principles that morally ought to undergird all research on human beings. The code contained the following ten principles, which seem to be by and large common-sense corollaries of democratic social consensus ethics.

1. The voluntary consent of the human subject is absolutely essential. This means that the person involved should have legal capacity to give consent; should be so situated as to be able to exercise free power of choice, without the intervention of any element of force, fraud, deceit, duress, over-reaching, or other ulterior form of constraint or coercion; and should have sufficient knowledge and comprehension of the elements of the subject matter involved as to enable him to make an understanding and enlightened decision. This latter element requires that before the acceptance of an affirmative decision by the experimental subject there should be made known to him the nature, duration, and purpose of the experiment; the method and means by which it is to be conducted; all inconveniences and hazards reasonably to be expected; and the effects upon his health or person which may possibly come from his participation in the experiment.

   The duty and responsibility for ascertaining the quality of the consent rests upon each individual who initiates, directs or engages in the experiment. It is a personal duty and responsibility which may not be delegated to another with impunity.
2. The experiment should be such as to yield fruitful results for the good of society, unprocurable by other methods or means of study, and not random and unnecessary in nature.
3. The experiment should be so designed and based on the results of animal experimentation and knowledge of the natural history of the disease or other problem under study that the anticipated results will justify the performance of the experiment.
4. The experiment should be so conducted as to avoid all unnecessary physical and mental suffering and injury.
5. No experiment should be conducted where there is an a priori reason to believe that death or disabling injury will occur; except, perhaps, in those experiments where the experimental physicians also serve as subjects.
6. The degree of risk to be taken should never exceed that determined by the humanitarian importance of the problem to be solved by the experiment.

7. Proper preparations should be made and adequate facilities provided to protect the experimental subject against even remote possibilities of injury, disability, or death.
8. The experiment should be conducted only by scientifically qualified persons. The highest degree of skill and care should be required through all stages of the experiment of those who conduct or engage in the experiment.
9. During the course of the experiment the human subject should be at liberty to bring the experiment to an end if he has reached the physical or mental state where continuation of the experiment seems to him to be impossible.
10. During the course of the experiment the scientist in charge must be prepared to terminate the experiment at any stage, if he has probable cause to believe, in the exercise of the good faith, superior skill and careful judgment, required of him that a continuation of the experiment is likely to result in injury, disability, or death to the experimental subject.[11]

Though the Nuremberg Code has been hailed as a beacon, it has also been widely criticized as inadequate and too abstract to regulate all manner of research on human beings. Though many people rightly feel that informed consent is a key concept for ethical research on people, it is not sufficient. As one doctor/researcher told me, chillingly: "I'm a really nice guy. I can convince anyone to consent to any kind of research." What counts as "informed"? Can desperate people really freely give consent? Can children? At what age? And so on.

The Declaration of Helsinki, first proposed by the World Medical Association in 1964 and regularly modified since then, attempted to supplement the Nuremburg Code in three major ways:[12] By distinguishing between research that benefits society rather than the subject (or between therapeutic clinical research and basic biomedical research), by establishing an institutional review committee to monitor ethical compliance, and by allowing proxy consent by family or guardians for subjects such as children or those cognitively compromised.

The reception of the Nuremberg Code by the U.S. research community and the correlative need for regulation of research on humans, as Dr. Jay Katz has remarked, "was dismissed out of the belief that the

[11] Http://ohsr.od.nih.gov/nuremberg.php.
[12] Brody, *Ethics of Biomedical Research*, p. 34.

concentration camp experiments were solely the results of the 'ravages of Nazi pseudoscience.'"[13] The U.S. research community saw no need for regulation and held firmly to the belief that professional judgment was the best protection for research subjects. "Academic freedom" and the fear of stifling research were valued far more highly than assuring ethical treatment of subjects.[14] Interestingly, the United States did not sign the Declaration of Helsinki.

By and large, the research community in the United States did not take seriously the ethics of using human subjects and laid blame for revelations of atrocities at the feet of the Nazis. The issues in the United States, though very much present, were essentially invisible until the 1960s, when they were dramatically revealed to the public both by close media scrutiny and by scholarly attention. Until then, the idea of regulation was anathema to the research community, and the antiregulatory complacency was typified by the statement that "the research subject has no better protection than the conscience of the principal investigator."

Strangely enough, the same sentiment is echoed by the scientist most responsible for alerting both the research community and the general public in the United States, Henry Beecher.[15] In a seminal article published in the *New England Journal of Medicine* in 1966 (after many rejections and demands for revision), Beecher pointed out that funding for research had increased exponentially and that, by searching the literature, he had found numerous examples of ethical problems in experimental medicine, and "in many of the examples presented, the investigators have risked the health or life of their subjects."[16] Of 100 human studies published in one journal in 1964, he found twelve that seemed to be unethical.[17] Of fifty studies he originally compiled, only two mentioned consent. In the article, for "reasons of space" he summarized twenty-two morally problematic studies.[18] These included withholding of a known treatment (penicillin) for streptococcal respiratory infection, so that those who

[13] Katz, "Regulation of Human Experimentation in the United States."
[14] Ibid.
[15] Beecher, "Ethics and Clinical Research."
[16] Ibid.
[17] Ibid.
[18] Ibid.

did not get the treatment developed rheumatic fever, and two other similar denials of proven therapy; drug therapy known to produce liver damage deliberately given to children to study the damage; drug doses known to produce aplastic anemia given to patients; anesthetized patients given enough CO_2 to produce cardiac arrhythmias; cerebral circulatory insufficiency induced in patients; and melanoma transplanted from daughter to mother, who subsequently died of metastatic disease.

After delineating these cases, Beecher suggests some solutions. He feels that journals should avoid publishing data from such experiments, though he admits to seeing the argument against this. Second, one should always strive for informed consent, though he is doubtful that this will do much good – in part because risks may not be clearly known, in part because physicians are so trusted that patients willingly accede to their requests, and in part because it is difficult to know what the patient comprehends. In the end, he affirms with the trend of his time that "a far more dependable safeguard than consent is the presence of a truly *responsible* investigator," a point he both begins and concludes with in his essay: "The more reliable safeguard [than consent] is the presence of an intelligent, informed, conscientious, compassionate, responsible investigator."[19]

However noble in intention, this is vacuous. In the first place, it comes perilously close to being a tautology – "the best way to have ethical research is to have ethical people do the research." Second, no mechanism is suggested for creating or screening for such ethical researchers.

After Beecher wrote, it has gradually become clear that he only saw the tip of a vast iceberg. Scholars, journalists, lawyers, and victims have revealed that abuse of human subjects was rife and epidemic for much if not most of the twentieth century, before and after Beecher, both in the United States and in other Western democratic societies lacking the excuse of Hitlerian ideology. Such abuse occurred in medical research, behavioral research, military research, and nutritional research, in fact, in all areas of research involving human subjects and to all manners of subjects – soldiers, patients, students, children, and entire urban and rural populations.

[19] Ibid.

To chronicle these abuses would take multiple volumes. For our purposes, it suffices to provide a sampling of the variety and scope of such atrocities. One of the best sources, published shortly after Beecher's paper in 1967, was a book entitled *Human Guinea Pigs* by a British physician named D. H. Pappworth.[20] Although virtually unknown then and now (Beecher mentions him once as a personal correspondent; none of my colleagues in bioethics has heard of him), Pappworth had valiantly tried to awaken medical and public awareness of morally bankrupt research, back into the 1950s, only to be censored and rebuked by his peers and largely ignored, since he was not as prominent as Beecher. In *Human Guinea Pigs*, he provides over five hundred examples of questionable research gathered over more than twenty years, primarily in the United States and Britain, arguably among the world's most enlightened countries. These examples are chilling and were not hidden; they were all taken from published reports in medical journals. Pappworth also remarks, "It must also be emphasized that many experiments are never reported in medical journals, and I have good grounds for believing that they are often the worst of all."[21]

Pappworth divides his discussion of questionable experimentation into thirteen categories, whose titles alone stir the dark side of our imagination. For brevity, I will list some examples from some of his chapters as well as other sources.

1) Experiments on Infants and Children

In order to establish the normal values for cerebral blood flow in children the following experiment on nine children aged three to eleven was carried out. No other details about the children are given except the fact that they did not have brain disease. Twelve adults, about whom no details are recorded, were used as controls.

First, a needle was inserted into the femoral artery (the main artery of the thigh). Second, the jugular vein (the main vein of the neck) was punctured deeply by another needle just below the angle of the jaw so that this needle could penetrate to the bulb of this vein. Third, while these two needles were in position the patient was made to inhale a special gas mixture through a tightly fitting face mask.

[20] Pappworth, *Human Guinea Pigs.*

[21] Ibid., p. 5.

I quote the above case because, apart from the unpleasantness of the procedure, the experimenters themselves underline what they were doing when they make four observations. The first is the note that "to minimize the pain in obtaining blood samples very sharp needles were used." The second is that two three-year-olds were among those submitted to the experiment who "required some restraint." The third is that the children's legs were immobilized by bandaging them to a board. The fourth is that "*thirty-five others* were originally selected for the procedure, but at one stage or another they became uncooperative."[22]

2) Experiments on Pregnant Women

The following experiment graphically illustrates the ethical obtuseness of major journals, in this case the *American Journal of Obstetrics and Gynecology*:

A group of Cuban doctors not only gave spinal anesthetics to a group of pregnant women but then proceeded to perform translumbar aortography on them. This was the first time that translumbar aortography had ever been done on pregnant women and its purpose was to study the abdominal circulation in advanced pregnancy.

After the spinal anaesthetic, with consequent paralysis and loss of sensation of the lower limbs and trunk, the patients were turned to lie face downwards, a six-inch needle was then pushed through the mass of muscles adjacent to the vertebral column until it entered the aorta, the main artery of the body. When blood flowed in "a pulsating stream" a contrast medium was injected through the needle and serial X-ray films were taken of the aorta and the renal arteries. The experiment was done on twelve women during the eighth month of pregnancy. One woman died of meningeal hemorrhage following the lumbar puncture given for the spinal anaesthetic.

The editor of the journal in which this experiment was reported appended a remarkable note to the article:

This paper, contrary to our custom of not accepting contributions from foreign sources, is presented because of the unusual character of the daring experiment.

The word "daring," which is certainly not out of place, signifies the taking of a major risk. But who ran this risk? Was it the patient, the doctors, or the unborn infant?[23]

[22] Ibid., p. 38.
[23] Ibid., pp. 47–48.

3) Experiments on Mental Defectives and the Mentally Sick

An extensive trial of an experimental measles vaccine was conducted in England and for this purpose fifty-six mentally subnormal children were selected. The authors state, "They were especially suitable for the study since close medical supervision was possible throughout." Twenty-one other mental defectives were not vaccinated, but were kept in close contact with the vaccinated group.

Forty-six of the vaccinated became feverish and forty-eight developed a rash. Twenty-two had a moderately severe and nine a very severe reaction to the vaccination. It is noted,

Many of the vaccinated became miserable and fretful during the period of rash and pyrexia.

Five of the children developed tonsillitis as a consequence and one had a severe complicating broncho-pneumonia.[24]

Pappworth goes on to quote Nobel Prize winner Alexis Carrel of the Rockefeller Institute arguing in favor of Hitler's extermination policy toward the mentally defective. *60 Minutes* recently reported normal children being "dumped" into a horribly repressive "home" for the feeble-minded in Massachusetts, where researchers from MIT did radiation experiments on them without consent into the 1960s.

4) Experiments with Prison Inmates

The vulnerability of prisoners to coercion made them perfect research subjects. In 1963, *Time* magazine cited examples of prisoners being injected with live cancer cells to see whether cancer would be produced. Another prisoner received an experimental lung transplant.[25]

A friend of mine who did painful dental research the 1960s told me that when he and other dental researchers needed human subjects for painful experiments he "just went to some Southern prisons with lots of cartons of cigarettes" to recruit "volunteers."

5) Experiments on the Dying and the Old

In 1961 an experimental technique was tried out by which it was hoped to show enlarged glands in the chest. For this investigation fifty-one patients were chosen who had cancer, two others who had heart failure and a further

[24] Ibid., p. 57.
[25] Ibid., p. 65.

three who had cirrhosis of the liver. The doctors also used eleven "normal controls" about whom no details are given.

Under local anesthetic a needle was inserted into a rib and "slowly and carefully driven with a small mallet" until the needle entered the bone marrow. When this had been done a contrast medium was rapidly injected through the needle and serial X-ray photographs were taken to outline the azygos veins of the chest, and possibly thereby to show up any enlarged glands.[26]

6) Experiments on Nonpatient Volunteers

Here Pappworth discusses primarily students. The issue here, of course, is overt or subtle coercion. When I was a college student in the 1960s, one could not pass the Introduction to Psychology course unless one served as a research subject in an experiment, which experiments were sometimes painful. At various institutions in the 1960s, graduate students served as subjects in nutrition experiments conducted by their advisers. Here is one of Pappworth's examples:

The following account was given by a postgraduate student of mine who witnessed in America a highly complex type of cardiac catheterization. The subject was a hard-up student who was paid for volunteering and had volunteered and been paid for similar experiments previously. During the present experiment the young man developed profound shock and collapse, followed by stoppage of the heart. He was successfully brought back to life, having, in fact, for a few moments been virtually dead. The experimenter then *continued with the experiment as though nothing had happened.* The experimenter's true attitude to his subject seems to have been expressed in the remark he made to those present at his performance: "He must be a fool to repeatedly come back to us."[27]

7) Experiments on Patients Awaiting Operations and as Extensions of Operations

The following experiment is one of which I have read the published account several times with increasing amazement. The subjects were eighteen patients described as undergoing either "injection type lobotomy" (which is otherwise known as leucotomy and is a purposeful destruction of part of the white matter of the brain in an attempt to relieve mental symptoms); or "cervico-thoracic

[26] Ibid., p. 77.
[27] Ibid., p. 86.

sympathectomy" (a cutting of the sympathetic nerve in the neck and chest to relieve a variety of symptoms, including intractable pain); "or other head and neck operations" not specified. None of the patients had any heart disease or any neurological disease other than mental. That is to say, none of them was suffering from a condition which the results of the experiment were aimed at illuminating. The purpose of the experiment was, in fact, to record the effects of various procedures on brain circulation. The experiment was carried out prior to operation.

First, a small hole (known as a trephine) was made in the temporal region of the skull under local anaesthetic – presumably after the head had been partly or wholly shaved – and the membrane covering the brain was "widely opened." Next, a "trephine button" was inserted into the skull through the trephine hole. From this "button," tubes lead to instruments which recorded cerebral circulation. This recording was done to show the effects of various procedures, namely, compression of the main veins of the neck; inhalation of a toxic gas (carbon dioxide); voluntary holding the breath for a considerable period; the intravenous injection of various substances, including nicotinic acid and also a half a pint of ten percent solution of alcohol. (This last I would consider itself a dangerous procedure.)

This, however, was not all. The experimenters record that "Apprehension was produced by threatening the patients with immersion of the left leg in iced water." In some patients the experimenters recorded the effects on brain circulation of "cutaneous pain." (Two of the patients had abdominal cancer.) The experimenters comment, "The trephine button may be left *in situ* (in place) indefinitely, avoiding the need to perform further surgery to remove the button." And they add, "It is possible to take records for up to twenty-three days."[28]

8) Radiation Exposure Experiments

We discuss this below in the context of the investigation of Department of Energy studies in the 1990s.

9) The Use of Patients as Controls

Pappworth explains that doctors in research institutions feel that they "own" their patients and offer them as controls in studies irrelevant to their illnesses, but involving deleterious treatments.[29]

[28] Ibid., pp. 87–88.
[29] Ibid., p. 103.

10) The Inducement of Illness in Subjects

Known toxic nitrogenous substances were given to eighteen patients who had cirrhosis of the liver. Of these patients five had persistent and seven transient neuro-psychological disorders. The other seven were in the terminal stages of the disease and died within eight days of the study. Eleven of the eighteen were also submitted to hepatic vein catheterization. The article reporting this experiment begins, "Nervous disorders have been induced in patients with liver disease by administration of certain nitrogenous substances." And the same article concludes, "Exacerbation followed a diet containing more than 70 gramme protein (the main source of nitrogen), or containing added methionine (another nitrogenous substance), or ammonium chloride."[30]

11) Experiments on Patients with Heart Disease

12) Research on Kidney Diseases

13) Experiments in Which New Drugs Are Tested

As detailed, well informed, and ethically sound as Pappworth's work was, it attracted little attention and is virtually forgotten today. And just as Beecher indicated that his paper went through much editorial attention before it was finally published, Pappworth indicates that even his numerous letters to the editors of various medical journals were rejected or heavily edited, and finally ignored by the research community.

Pappworth's work covered research from roughly the 1940s through the mid-1960s. Unfortunately, other scholars have documented the same pattern earlier in the twentieth century. Two excellent books are available covering questionable aspects of research on human beings during that period: Susan Lederer's *Subjected to Science: Human Experimentation in America before the Second World War* (1999) and the early chapters of Jonathan Moreno's *Undue Risk: Secret State Experiments on Humans* (1995). The pattern unearthed by these scholars is totally congruent with Pappworth's results.[31] Moreno's book points out that many of the morally questionable projects undertaken during the twentieth century were government-sponsored and were done

[30] Ibid., p. 133.
[31] Http://tis.eh.doe.gov/ohre.

under various versions of the rubric of national defense. In particular, Moreno shows a pattern of using military personnel, without proper informed consent, going back to Walter Reed's experiments on yellow fever in 1900 and persisting through Operation Desert Storm.

Experiments sponsored by the military and/or using military personnel represent a gruesome litany. During the 1940s, servicemen and conscientious objectors were used for mustard gas studies. In 1945, the military and the State Department initiated Operation Paper Clip, giving safe haven to Nazi scientists. In 1946, patients in Veterans Administration hospitals began to be used in medical experimentation. In 1947 the CIA began LSD experiments using subjects both with and without consent. (The 1990 film *Jacob's Ladder* is based on these experiments with soldiers.) In 1950, the Department of Defense began to monitor morbidity and mortality of humans in desert areas where nuclear bombs were tested. Also in 1950, studies began of the degree to which cities could be incapacitated by release of pathogens. In 1966, bacteria were released in the New York subway system after years of other such trials. (In 1977, Senate hearings confirmed that 239 populated areas had been deliberately contaminated with biological agents between 1949 and 1968.) Mind and behavior control experiments were prominent through the 1960s, conducted by the military in the United States, Europe, and Asia. In 1994, Senator John D. Rockefeller issued a report detailing fifty years of illicit experiments on hundreds of thousands of military personnel. In 1995, the government admitted to courting Japanese war criminals who had performed outrageous experiments and offering them immunity in exchange for working on biological warfare.

Also in 1995, the Department of Energy published its history of human radiation experiments, documenting over 400 experiments illicitly performed by the Department of Energy (DOE), the Atomic Energy Commission (AEC), and their predecessors on human beings, including civilians, children, radiation workers, populations, uranium mine workers, and soldiers. The records documenting these stories occupy 3.2 *million* cubic feet of space. The experiments are summarized in a massive DOE Web site entitled Human Radiation Experiments[32] and are well discussed in Moreno's book.

[32] Moreno, *Undue Risk*, p. 279.

In the face of this staggering and appalling record of immoral research, it is difficult to pinpoint the areas that most awakened public concern. Nonetheless, there are three incidents that stand out in terms of the shock value they created.

Probably most dramatic was the Tuskegee syphilis study, an incident that Moreno has rightly called "the most notorious medical research project in American history."[33] The study ran from 1932 to 1972, during which time four hundred poor and uneducated black sharecroppers who had syphilis were observed and tested while not being told the nature of their affliction. They were subjected to repeated invasive tests such as spinal taps and never received current state-of-the art therapy nor penicillin when it became available as a cure. In an era highly conscious of civil rights, this case represents intolerable behavior by the medical community (the Public Health Service).[34] Significantly, although the studies were published over forty years ago, no one in the syphilis research community raised objections, and the Public Health Service director defended the "ethics" of the studies when they were exposed in 1972.

The second case, revealed in 1963, occurred at the Jewish Chronic Disease Hospital in Brooklyn, New York, where doctors injected live cancer cells into twenty-two chronically ill patients in a regimen totally unrelated to their therapy, without obtaining consent.[35]

The third case took place in New York at the Willowbrook State School for mentally retarded children. These studies, which were conducted over fifteen years, were designed to develop a vaccine for hepatitis and involved deliberately exposing the children to the disease. The conditions were horrific at Willowbrook, and although parents had given consent, such consent was presuppositional to admitting the children to the school.[36]

[33] For a detailed discussion of this study, see the CDC Tuskegee Syphilis Study page at http://www.cdc.gov/nchstp/od/tuskegee/index.html, accessed October 8, 2005, and the Syphilis Study Legacy Committee Final Report, May 20, 1996, at www.med.virginia.edu/hs-library/historical/apology/report.html, accessed October 8, 2005.

[34] Katz, *Experimentation with Human Beings*, pp. 9–65.

[35] Ibid.

[36] See "The Willowbrook Letters," from the *Lancet*, reprinted in Ronald Munson (ed.), *Intervention and Reflection: Basic Issues in Medical Ethics*, 6th ed. (Belmont, Calif.: Wadsworth, 1994), pp. 508ff.

How can we even begin to comprehend the prevalence of the patently unethical research we have chronicled? In many ways, it is easier to understand the Nazis – at least they were ideologically brainwashed. But how could *we* do such things? (Ironically, lawyers for the Nazi researchers invoked the history of U.S. experimentation as a defense, a defense that crumbled only when a senior U.S. medical researcher lied about U.S. policy.)[37]

A clue as to how this could occur can be found in a remarkable article by Dr. Jay Katz, a physician and attorney who has been a major figure in the ethics of research on human beings in the United States and is himself a Holocaust survivor. The article, entitled "The Regulation of Human Experimentation in the United States – A Personal Odyssey," was published in the journal *Institutional Review Board* in 1987. In this article Katz chronicles his own involvement with human research regulation, and begins with this anecdote:

> During 1954 to 1958, two colleagues and I conducted experiments on hypnotic dreams, supported by grants from the National Institutes of Health. After we had worked out our research methodology, we became concerned that some of our volunteer-subjects, as a consequence of their participation, might experience sufficient emotional stress to require a brief period of hospitalization. While we believed this to be unlikely, we were sufficiently worried to ask the chairman of our department whether, if our fears materialized, he would make hospital beds available to our subjects free of charge. He readily agreed. *We were relieved, feeling that we had satisfactorily fulfilled our professional responsibilities to our research subjects. It did not even occur to us to wonder: Were we obligated to disclose to our subjects our concerns about the investigation's possible detrimental impact?* [emphasis added][38]

"It did not even occur to us to wonder." That sentence is a mark of an ideological blindfold, eclipsing ordinary common decency. No ordinary person of common sense would fail to think about disclosure. But even Katz, a highly sensitive individual, was so blinded by scientific ideology and its claims about value-free science that it never occurred to him to question the fundamental use of a person to achieve a scientific goal. His ethical nature shone through only on the relatively small point of financial responsibility for hospitalization. In other words, in

[37] See Moreno, *Undue Risk*, chapter 3.
[38] Katz, "Regulation of Human Experimentation," pp. 1–6.

its own way, what we have called scientific ideology or scientific common sense thoroughly concealed the key ethical question of using a human being – obtaining informed consent – by totally submerging the dictates of ordinary common sense and ordinary common decency.

What follows in Katz's article is a remarkable confession:

> Years later, I learned from my law colleagues and students much about citizens' rights to self-determination and privacy, and the limits of governmental and professional intrusions to persons' private lives. These concerns, it dawned on me, had relevance to my own past research, to the interactions between physician-investigators and patient-subjects, and to the controls the medical profession must impose on the research process. They had not been explored in my medical education; I had not considered them when I invited my research subjects to join me in my projects. Although I was quite disturbed for a while about my thoughtlessness and the lacunae in my education, I soon realized that the problem of human experimentation cannot be adequately resolved, as generations of doctors had been taught, by a pledge of allegiance to such undefined principles as *primum non nocere* or to visionary codes.[39]

In other words, only by becoming radically immersed in a different, ethically charged career (law), could Katz begin to escape the shackles of the ideology he had acquired in his science and medical education. The chilling implication is that even a man like Katz would not have realized the ethical underpinnings and presuppositions of his research activity had he not had the opportunity to immerse himself in another field carrying a contrary set of ideological commitments.

Recalling our earlier discussion of theories and expectations conditioning perception and recalling that an ideology *is* an overarching, pervasive theory that filters one's thinking and experience, we may conclude that immersion in scientific ideology creates people blinded to ethical concerns.

Medical training, I believe, is particularly conducive to the imposition of scientific ideology. Over my many years of sending premedical students to medical school and teaching veterinary students, I have watched with alarm how too many of them, even after a year or two of medical education, began to carry themselves like doctors. They become remote, aloof, detached from the sentiments, emotions, and moral commitments that drove them to medicine as a career in the

[39] Ibid., 1–2.

first place. To some extent this is necessary, required both to keep their sanity (excessive empathy would make medical life impossible to carry on) and to solidify their Aesculapian authority, the powerful authority that inheres in a medical professional in virtue of his or her profession. No one could yell at Hitler except his doctor; no one can probe you as painfully and intimately as a doctor. But besides these legitimate forces, there are others: the "God complex," the belief that nonphysicians can't understand the pressures physicians are under, the intentional building up of a strong "fraternity" mentality, and the sense of superiority that comes with dealing with matters of life and death.

There is another hint in Beecher explaining how researchers can forget or be blinded to basic ethical principles of thinking of human beings as ends in themselves, having respect for individuals, treating them with justice and fairness, receiving informed consent, and so on, that should be apparent to common sense. I am referring to his brief comment at the end of this article.

An experiment is ethical or not at its inception. It does not become ethical *post hoc* – ends do not justify means.[40]

When one is trained in science, one is taught that garnering knowledge is the end of science, its goal. Knowledge is seen as intrinsically good (though it can be wrongly deployed). The highest good for a scientist is to achieve knowledge that will eventuate in "our" ability to control nature and make a better world, eliminate disease, and so on. This conviction is almost a religious belief in its unshakeability. Philosopher Robert Paul Wolff once facetiously characterized this attitude by saying that each scientist feels that he or she is responsible for throwing a little piece of dung into a huge dung-heap, and eventually there will stand a cathedral.

So we can augment what we said about researcher ideology in the following way: "What we are doing is noble and valuable and will create vast benefit for humanity. Ethical questions are not our purview, since science is value-free. But, in any case, our results more than justify the steps we take to achieve them."

40 Beecher, "Ethics and Clinical Research," p. 60.

I am reminded of an old French film with the actor Fernandel, where he plays a man who may or may not be God, who is asked by a farmer how he can allow the innocent to suffer. His response: "Do you think of the ants you destroy when you plow the field?" Something similar seems to be operative here.

This, of course, is a bad "argument." As we saw in discussing ethics, utilitarianism – greatest good for the greatest number – is only part of our social ethic. It must always be balanced by Kantian respect for individuals. This is, in essence, widely known as "common sense" to people not involved with scientific ideology – that is why ordinary people are so shocked at the atrocities we chronicled. To ordinary moral consciousness, virtually no benefit justifies hurting innocents. (I say virtually because saving the world might count as a reason, but certainly none of the standard experiments we chronicled even begins to approach that degree of utilitarian value.)

To these considerations explaining the shocking immorality – or amorality – of much twentieth-century use of human subjects, we can add some others that help to explain (but not to justify) such use in a few areas. With regard to the use of military personnel, there may well have been a presumption among researchers that such people are *expected* to risk their lives in military service on the battlefield. Since we were, as it were, at war with disease and constantly in a cold war state, it was legitimate to use these people as needed, as a logical extension of their being "soldiers."

In the case of prisoners, there was (and probably still is) a widespread presumption in society that such people are *guilty*. Thus, using them in painful or dangerous experiments is a way of their *atoning* or paying their debt to society. To this day, many of my students find experimenting on "guilty" prisoners less morally problematic than experimenting on "innocent" animals.

Finally, there is no question that retarded children and adults, the insane, and black people were considered inferior for much of the twentieth century. It is probably only the post–civil rights generation of Americans who have not been exposed to that view.

It is also fascinating that when some of the major human-use scandals were made public, many members of the scientific community supported the questionable research. Robert Levine, in his *Ethics and*

Regulation of Clinical Research,[41] quotes a 1975 editorial by the editor-in-chief of the *New England Journal of Medicine*, Franz Ingelfinger, where he states that

> reports of investigations performed unethically are not accepted for publication. Thus, appearance of another Willowbrook report in this issue indicates that the study on balance, is not rated as unethical.

In any case, by the 1960s, Congress was looking at research on human beings with an eye to creative legislation. In 1966, the NIH/Public Health Service, fearful of research's being stifled by congressional action, announced a policy of self-regulation to be administered by local review committees with loss of government funding as a sanction for noncompliance. This process did not really begin to be operational until the 1970s.

According to Jay Katz, the concept of review boards was doomed from the start:

> [NIH] rejected the idea of establishing substantive guidelines for the conduct of research... placing its faith in the judgment of investigators' peers. This decision ducked all the important issues the Council had to consider.... The idea of establishing institutional review committees had merit, but not as the major corrective for all the concerns over the conduct of research.[42]

Katz goes on to quote approvingly from a nonphysician present at the discussions chartering institutional review boards (IRBs):

> He was dumbfounded by the idea of institutional review committees that delegated all responsibility to them, when at the same time the ethical conduct of research seemed in such disarray. He said, "You cannot tell people to do the right thing without telling them what the right thing is." He also warned that... the regulations would eventually become more detailed and extensive. This is exactly what happened.[43]

History has indeed shown that the IRB system was inadequate, but not, I believe, for the reasons Katz cites. Consider, for a moment, the claim just cited that scientists were being asked to serve as a jury without knowing the law. *I would argue, on the contrary, that any reasonably educated*

[41] Levine, *Ethics and Regulation of Clinical Research*, p. 71.
[42] Katz, "Regulation of Human Experimentation," p. 3.
[43] Ibid.

citizen – even of high school age – should be sufficiently educated in our social consensus ethic to be able to extrapolate that ethic to the issue of ethical treatment of research subjects. Anyone who has had even a high school civics class, after all, knows that we do not oppress the individual *even* for the general welfare, but respect individual rights. I would guess that, if we were to ask such a high school student about giving LSD to soldiers without informing them, piggybacking an experimental surgery on top of a caesarian section, or indeed about any of the cases we cited earlier, they would be pretty clear on what our ethic dictated. Even one of the most recent U.S. deaths in an experiment, that of sixteen-year-old Jesse Gelsinger in the trial of a gene therapy modality for clearing ammonia, which occurred at the University of Pennsylvania, would be readily seen as morally unacceptable on the grounds that the subject had not been fully informed of risks that emerged in animal experiments and that his blood ammonia levels exceeded what the protocol itself deemed acceptable. (Indeed, many ordinary people with whom I have discussed the case do not hesitate to call it "murder.")

In my view, the issue is not ignorance of ethics, but rather the ideological carapace that insulates people saturated with scientific ideology from thinking about ethics in the context of science, as we have discussed in preceding chapters. To mandate the application of ethics by a peer group to the analysis of proposed scientific research is fine for people who believe in the relevance of ethics to science, but it is meaningless for people who believe science or medicine is ethics- or value-free.

In a perfectly well-intentioned way, the federal government's Department of Health, Education and Welfare went on, in 1979, to produce the Belmont report, which has served as the moral basis for federal regulation in ensuing years.[44] According to Belmont, research on humans must be guided by three principles: respect for persons, beneficence, and justice.

Respect for persons, according to the report, incorporates at least two ethical convictions:

> First, that individuals should be treated as autonomous agents, and second, that persons with diminished autonomy are entitled to protection.

[44] Full text can be found at http://ohsr.od.nih.gov/guidelines/belmont.html, accessed October 8, 2005.

From the latter claim, of course, came the extensive regulations protecting children, prisoners, and so on. In fact, prisoners are cited as an example of people with diminished autonomy where such diminution may not be obvious.

The second principle, that of *beneficence*, meant two things:

> (1) do not harm; and (2) maximize possible benefits, and minimize possible harms. . . . The problem posed by these imperatives is to decide when it is justifiable to seek certain benefits despite the risks involved, and when the benefits should be foregone because of the risks.

Finally, the principle of *justice* is concerned with "who ought to receive the benefits of research and bear its burdens?" The text cites numerous examples of injustice as when, in the nineteenth and early twentieth centuries, poor ward patients served as research subjects, with the benefits of such research flowing largely to private (wealthy) patients. (Earlier in the nineteenth century, the same point could have been made about slaves.) Other examples of injustice includes the Nazi concentration camp prisoners and the Tuskegee disadvantaged, rural, uneducated black men who were deprived of effective treatment.

The application of these principles to human research is worth quoting *in toto*:

> Applications of the general principles to the conduct of research leads to consideration of the following requirements: informed consent, risk/benefit assessment, and the selection of subjects of research.
>
> 1. Informed Consent
>
> Respect for persons requires that subjects, to the degree that they are capable, be given the opportunity to choose what shall or shall not happen to them. This opportunity is provided, when adequate standards for informed consent are satisfied.
>
> While the importance of informed consent is unquestioned, controversy prevails over the nature and possibility of an informed consent. Nonetheless, there is widespread agreement that the consent process can be analyzed as containing three elements: information, comprehension and voluntariness.
>
> - **Information**
>
> Most codes of research establish specific items for disclosure, intended to assure that subjects are given sufficient information. These items generally

include: the research procedure, their purposes, risks and anticipated benefits, alternative procedures (where therapy is involved), and a statement offering the subject the opportunity to ask questions and to withdraw at any time from the research. Additional items have been proposed, including how subjects are selected, the person responsible for the research, etc.

However, a simple listing of items does not answer the question of what the standard should be for judging how much and what sort of information should be provided. One standard frequently invoked in medical practice, namely the information commonly provided by practitioners in the field or in the locale, is inadequate, since research takes place precisely when a common understanding does not exist. Another standard, currently popular in malpractice law, requires the practitioner to reveal the information that reasonable persons would wish to know in order to make a decision regarding their care. This, too, seems insufficient, since the research subject, being in essence a volunteer, may wish to know considerably more about risks gratuitously undertaken than do patients who deliver themselves into the hand of a clinician for needed care. It may be, that a standard of "the reasonable volunteer" should be proposed: the extent and nature of information should be such that persons, knowing that the procedure is neither necessary for their care nor perhaps fully understood, can decide whether they wish to participate in the furthering of knowledge. Even when some direct benefit to them is anticipated, the subjects should understand clearly the range of risk, and the voluntary nature of participation.

A special problem of consent arises, where informing subjects of some pertinent aspect of the research is likely to impair the validity of the research. In many cases, it is sufficient to indicate to subjects that they are being invited to participate in research, of which some features will not be revealed until the research is concluded. In all cases of research involving incomplete disclosure, such research is justified, only if it is clear that (1) incomplete disclosure is truly necessary to accomplish the goals of the research, (2) there are no undisclosed risks to subjects that are more than minimal, and (3) there is an adequate plan for debriefing subjects, when appropriate, and for dissemination of research results to them. Information about risks should never be withheld for the purpose of eliciting the cooperation of subjects, and truthful answers should always be given to direct questions about the research. Care should be taken to distinguish cases, in which disclosure would destroy or invalidate the research, from cases in which disclosure would simply inconvenience the investigator.

- **Comprehension**

The manner or context in which information is conveyed is as important as the information itself. For example, presenting information in a disorganized and rapid fashion, allowing too little time for consideration, or curtailing opportunities for questioning, all may adversely affect a subject's ability to make an informed choice.

Because the subject's ability to understand is a function of intelligence, rationality, maturity and language, it is necessary to adapt the presentation of the information to the subject's capacities. Investigators are responsible for ascertaining that the subject has comprehended the information. While there is always an obligation to ascertain that the information about risk to subjects is complete and adequately comprehended, when the risks are more serious, that obligation increases. On occasion, it may be suitable to give some oral or written tests of comprehension.

Special provision may need to be made, when comprehension is severely limited – for example, by conditions of immaturity or mental disability. Each class of subjects that one might consider as incompetent (e.g., infants and young children, mentally disabled patients, the terminally ill, and the comatose) should be considered on its own terms. Even for these persons, however, respect requires giving them the opportunity to choose, to the extent they are able, whether or not to participate in research. The objections of these subjects to involvement should be honored, unless the research entails providing them a therapy unavailable elsewhere. Respect for persons also requires seeking the permission of other parties in order to protect the subjects from harm. Such persons are thus respected, both by acknowledging their own wishes, and by the use of third parties to protect them from harm.

The third parties chosen should be those who are most likely to understand the incompetent subject's situation, and to act in that person's best interest. The person chosen to act on behalf of the subject should be given an opportunity to observe the research, as it proceeds, in order to be able to withdraw the subject from the research, if such action appears in the subject's best interest.

- **Voluntariness**

An agreement to participate in research constitutes a valid consent, only if voluntarily given. This element of informed consent requires conditions free of coercion and undue influence. Coercion occurs when an overt threat of harm is intentionally presented by one person to another, in order to obtain compliance. Undue influence, by contrast, occurs through an offer of an excessive, unwarranted, inappropriate or improper reward or other overture, in order to obtain compliance. Also, inducements that would ordinarily be acceptable may become undue influences, if the subject is especially vulnerable.

Unjustifiable pressures usually occur when persons in positions of authority or commanding influence – especially where possible sanctions are involved – urge a course of action for a subject. A continuum of such influencing factors exists, however, and it is impossible to state precisely, where justifiable persuasion ends and undue influence begins. But undue influence would include actions such as manipulating a person's choice through the controlling influence of a close relative, and threatening to withdraw health services to which an individual would otherwise be entitled.

2. Assessment of Risks and Benefits

The assessment of risks and benefits requires a careful arrayal of relevant data, including, in some cases, alternative ways of obtaining the benefits sought in the research. Thus, the assessment presents both an opportunity and a responsibility to gather systematic and comprehensive information about proposed research. For the investigator it is a means to examine whether the research is properly designed. For a review committee, it is a method for determining whether the risks that will be presented to subjects are justified. For prospective subjects, the assessment will assist the determination whether or not to participate.

- **The Nature and Scope of Risks and Benefits**

The requirement that research be justified on the basis of a favorable risk/benefit assessment, bears a close relation to the principle of beneficence, just as the moral requirement that informed consent be obtained is derived primarily from the principle of respect for persons.

The term "risk" refers to a possibility that harm may occur. However, when expressions such as "small risk" or "high risk" are used, they usually refer (often ambiguously) both to the chance (probability) of experiencing a harm, and the severity (magnitude) of the envisioned harm.

The term "benefit" is used in the research context to refer to something of positive value related to health or welfare. Unlike "risk," "benefit" is not a term that expresses probabilities. Risk is properly contrasted to probability of benefits, and benefits are properly contrasted with harms rather than risks of harm. Accordingly, so-called risk/benefit assessments are concerned with the probabilities and magnitudes of possible harms, and anticipated benefits. Many kinds of possible harms and benefits need to be taken into account. There are, for example, risks of psychological harm, physical harm, legal harm, social harm and economic harm, and the corresponding benefits. While the most likely types of harms to research subjects are those of psychological or physical pain or injury, other possible kinds should not be overlooked.

Risks and benefits of research may affect the individual subjects, the families of the individual subjects, and society at large (or special groups of subjects in society). Previous codes and Federal regulations have required that risks to subjects be outweighed by the sum of both the anticipated benefit to the subject, if any, and the anticipated benefit to society in the form of knowledge to be gained from the research. In balancing these different elements, the risks and benefits affecting the immediate research subject will normally carry special weight. On the other hand, interests, other than those of the subject, may on some occasions be sufficient by themselves to justify the risks involved in the research, so long as the subjects' rights have been protected. Beneficence thus requires that we protect against risk of harm to subjects, and also that we be concerned about the loss of the substantial benefits that might be gained from research.

- **The Systematic Assessment of Risks and Benefits**

It is commonly said that benefits and risks must be "balanced," and show to be "in a favorable ratio." The metaphorical character of these terms draws attention to the difficulty of making precise judgments. Only on rare occasions will quantitative techniques be available for the scrutiny of research protocols. However, the idea of systematic, nonarbitrary analysis of risks and benefits should be emulated insofar as possible. This ideal requires those making decisions about the justifiability of research to be thorough in the accumulation and assessment of information about all aspects of the research, and to consider alternatives systematically. This procedure renders the assessment of research more rigorous and precise, while making communication between review board members and investigators less subject to misinterpretation, misinformation and conflicting judgments. Thus, there should first be a determination of the validity of the presuppositions of the research; then the nature, probability and magnitude of risk should be distinguished, with as much clarity as possible. The method of ascertaining risks should be explicit, especially where there is no alternative to the use of such vague categories as small or slight risk. It should also be determined whether an investigator's estimates of the probability of harm or benefits are reasonable, as judged by known facts or other available studies.

Finally, assessment of the justifiability of research should reflect at least the following considerations: (i) Brutal or inhumane treatment of human subjects is never morally justified. (ii) Risks should be reduced to those necessary to achieve the research objective. It should be determined whether it is in fact necessary to use human subjects at all. Risk can perhaps never be entirely eliminated, but it can often be reduced by careful attention to alternative procedures. (iii) When research involves significant risk of serious impairment, review committees should be extraordinarily insistent on the justification of the risk (looking usually to the likelihood of benefit to the subject – or, in some rare cases, to the manifest voluntariness of the participation). (iv) When vulnerable populations are involved in research, the appropriateness of involving them should itself be demonstrated. A number of variables go into such judgments, including the nature and degree of risk, the condition of the particular population involved, and the nature and level of the anticipated benefits. (v) Relevant risks and benefits must be thoroughly arrayed in documents and procedures used in the informed consent process.

3. Selection of Subjects

Just as the principle of respect for persons finds expression in the requirements for consent, and the principle of beneficence in risk/benefit assessment, the principle of justice gives rise to moral requirements that there be fair procedures and outcomes in the selection of research subjects.

Justice is relevant to the selection of subjects of research at two levels: the social and the individual. Individual justice in the selection of subjects would require that researchers exhibit fairness: thus, they should not offer potentially beneficial research only to some patients, who are in their favor, or select only "undesirable" persons for risky research. Social justice requires that distinction be drawn between classes of subjects that ought, and ought not, to participate in any particular kind of research, based on the ability of members of that class to bear burdens, and on the appropriateness of placing further burdens on already burdened persons. Thus, it can be considered a matter of social justice, that there is an order of preference in the selection of classes of subjects (e.g., adults before children), and that some classes of potential subjects (e.g., the institutionalized mentally infirm or prisoners) may be involved as research subjects, if at all, only on certain conditions.

Injustice may appear in the selection of subjects, even if individual subjects are selected fairly by investigators, and treated fairly in the course of research. Thus, injustice arises from social, racial, sexual and cultural biases institutionalized in society. Thus, even if individual researchers are treating their research subjects fairly, and even if institutional review boards are taking care to assure that subjects are selected fairly within a particular institution, unjust social patterns may nevertheless appear in the overall distribution of the burdens and benefits of research. Although individual institutions or investigators may not be able to resolve a problem that is pervasive in their social setting, they can consider distributive justice in selecting research subjects.

Some populations, especially institutionalized ones, are already burdened in many ways by their infirmities and environments. When research is proposed that involves risks and does not include a therapeutic component, other less burdened classes of persons should be called upon first to accept these risks of research, except where the research is directly related to the specific conditions of the class involved. Also, even though public funds for research may often flow in the same directions as public funds for health care, it seems unfair that populations dependent on public health care constitute a pool of preferred research subjects, if more advantaged populations are likely to be the recipients of the benefits.

One special instance of injustice results from the involvement of vulnerable subjects. Certain groups, such as racial minorities, the economically disadvantaged, the very sick and the institutionalized, may continually be sought as research subjects, owing to their ready availability in settings where research is conducted. Given their dependent status and their frequently compromised capacity for free consent, they should be protected against the danger of being involved in research solely for administrative convenience, or because they are easy to manipulate as a result of their illness or socioeconomic condition.

In 1981, the federal government began to release regulations founded in the Belmont principles ethical framework. These regulations were revised and expanded in 1991 and 2001, continue to be revised, and became ever increasingly detailed and bureaucratic. (The reader can judge for him- or herself by perusing the Office for Human-Research Protections Web site at http://ohrp.asaphs.dhhs.gov.) In their most recent (2001) iteration, these regulations comprise thirty-four pages, single-spaced. For example, harsh new rules aimed at protecting patient privacy have made it incredibly difficult and bureaucratic to access patient records, prompting an editorial in the *New England Journal of Medicine* to opine, "The consequences for epidemiological health services and other public health research could be devastating."[45] Bear in mind that this covers issues relatively far removed from directly harming subjects.

Yet despite the governmental oversight, there is no reason to believe that things have improved; indeed, in the last few years we have seen continuing egregious infractions of ethical research on humans, despite a period of federal audits resulting in harsh penalties and loss of funding for delinquent institutions. For example, research deaths[46] based in numerous ethical violations of common decency and care for persons have occurred at the University of Rochester, Johns Hopkins University, the University of Pennsylvania, and Case Western Reserve University. Yet the government continues to bureaucratize regulations

[45] Kulnych and Korn, "Effect of the New Federal Medical Privacy Rule on Research," p. 1133.

[46] Stolberg, "The Biotech Death of Jesse Gelsinger," pp. 136–50.

and to make the procedure more draconian. Most extreme is the emerging suggestion that IRBs become policemen and spot-check on the actual conduct of research.

How can we explain the fact that while rules and penalties grow harsher, compliance with ethics in human research does not improve or grows worse? Obviously, one can only speculate, but I believe we can point to some plausible explanations.

In the first place, between the mid-1960s and the present, social thought has moved increasingly toward the individual-rights rather than utilitarian-benefit end of the spectrum of our social consensus ethic. This can be seen in many ways, from juries awarding millions to a woman who spilled hot coffee in her lap to general litigiousness to extreme political correctness wherein giving offense is the greatest sin to mega-heroic efforts protecting any historically disenfranchised group. Third, the inexorable social obsession with accountability makes ever-increasing scrutiny inevitable. All of these militate in favor of society's concern that subjects not be abused.

In my experience, IRBs in the mid-'70s right after their inception worked far better than they do today. The reason for this, I believe, harks back to the comment made by the person quoted in Katz's article regarding IRBs being like a jury without rules. In my view, this was an asset, not a detriment. In the absence of rules, IRB members were forced to think, to think ethically, and to recollect what they already knew of ethics. In a way, they were released from the blinders imposed by ideology, and they thought as ordinary citizens who, as I have argued, would when needed to do so be able to see readily many of the principles of medical ethics as corollaries of our consensus social ethic. The value of the very general rules was to force the sort of reflection that inevitably led to the reappropriation of common-sense ethics. In other words, the power of federal policy was not in providing details but in forcing reflection incompatible with scientific ideology and congenial to thinking in ordinary ethical terms.

During the mid-1970s, serving on the Colorado State University IRB, I watched researchers engage in dialogue to rediscover relevant ethical principles. In those days, the government was unconcerned about differences between institutions; it was taken for granted that institutional cultures would differ in ethical interpretations within the general framework of social consensus ethics. Indeed, IRB members from "different cultures" would meet at scientific gatherings and, as it were, cross pollinate. And as membership of IRBs changed, new cultural vectors were introduced. For example, at my institution (admittedly lacking a medical school), we were highly protective of children in behavioral research, often rejecting protocols that had been done with no problem at other institutions.

In short, the system at that time was conducive to dialogue and reflective consensus. Detailed bureaucratic rules dictating decisions were few, and ethical subtleties flourished. Tellingly, the office coordinating the IRB was then entitled "Program Ethics."

But as federal intrusion increased, ethical reflection decreased, and ideology again slipped into place. Instead of ethics becoming the focus of IRBs, meeting (or evading) regulations became the name of the game. More forms, more administrators, more hoops to jump through and barriers to leap replaced moral discussion. What leaps to mind is the old Aesop fable about the competition between the sun and the wind to get a traveler to remove his coat. The harder the wind would blow, the more tightly the man would pull his coat shut. The bureaucratic winds put the emphasis far more on regulation than on ethics, so the brief period wherein researchers were thrown back on their ethical resources and thus on recollection was truncated. More and more biomedical scientists saw the regulations as politically motivated, bureaucratic, politically correct rules, not as reflections of serious social ethics. Behavioral researchers saw these regulations as intended for medical researchers and irrelevant to them, though over the years behavioral work had drawn its share of flak – witness the Milgram experiments showing that most people would electro-shock a subject to heart attack if told to do so by an authority figure; the Stanford experiments where some students played the role of prisoners while others played jailers, and appalling cruelty emerged in the jailers; and various anthropologist intrusions in native cultures. Tellingly, our office in charge of IRBs had its name changed to "Regulatory Compliance," a nomenclature seemingly designed to cause eyes to glaze and jaws to set.

The last twenty years have seen the regulations flip-flop between forbidding pregnant women to participate in research that might harm their unborn child to regulations that guarantee them the right to participate. Similarly, with children as subjects, researchers used to exclude them to protect them, but now must include them unless they give a good reason. On a more amusing level, the IRBs once received a memo from Washington ordering us to call Indians "Native Americans," as "Indians" was demeaning. Barely six months later, we received a "never mind" memo. Someone had actually *asked* the Indians; they felt "Native" was demeaning, evoking "the natives are restless" and tom-tom beating, so we were to return to "Indians." Added to scientific ideology, then, was the belief that the regulations represent yet another example of political correctness gone wild. We are currently being pressed to have minority membership in the IRB in proportion to the minorities in the general population and to include alcoholics and drug addicts on committees to scrutinize protocols using alcoholics and drug addicts.

As the regulations and edicts grow ever more distant from common sense and common decency, researcher skepticism about ethics is overlaid with cynicism. Thought processes are so shallow that when hearing of a bioethicist on an IRB, a dean recently asked an IRB administrator in all innocence, "What does ethics have to do with research on humans?"

In addition to growing hostility between IRBs and the researchers they are supposed to represent, IRBs are underfunded and understaffed in a time of shrinking resources. This can lead to ill-considered or slow processing of protocol forms that further erodes committee credibility. If IRBs do begin to police research, the relationship between them and researchers will become strained to the breaking point. Since science tends to be ahistorical, as well as aphilosophical and "value-free," the historical reasons for IRBs are rarely appreciated.

In my view, if the government promulgates more and increasingly draconian regulation, this will have the same effect as the wind blowing harder in Aesop's story. Researchers are already disrespectful of the system, and look for ways to avoid and evade. There is only one way around this impasse.

I have already argued that scientific ideology trains scientists to avoid ethical and conceptual questions. This makes for bad ethics *and* bad science, for a whole range of inquiry is ruled out by fiat, and science is distanced from ordinary common sense. I would therefore suggest that all graduate education in science include a mandatory philosophical and ethical component of at least one year in duration, as well as a historical depiction of past wrongs. This, as my own experience attests, could gradually break the stranglehold of "science is value-free" ideology on the scientific community and lead scientists to engage the ethical issues whose denial and ignoring has led to an inestimable amount of loss of public credibility as well as to outright rejection of scientific advances by the public for fear that these advances are immoral (*vide* cloning, genetic engineering, and other biotechnological modalities). As we see below in the discussion of biotechnology, there is good evidence that ethical concern provides a greater social source of rejection of biotechnology than does fear; yet the regulatory and scientific communications and policies focus on fear and dangers.

I have seen a salubrious change in the moral acuity of young scientists since the NIH mandated courses in science and ethics for nascent scientists receiving training grants. Expanding this promising approach seems to me the only way to stop the constant exploitation of human subjects that we have witnessed for more than a century. Thinking in ethical terms must be made part of scientific thinking, and this can be achieved only by education.

5

Animal Research

If human subjects were treated as badly and cavalierly as we have documented throughout the twentieth century, one can anticipate that, a fortiori, animals used in research or science education had little chance of proper treatment. In this area, also, scientific ideology militated against scientists even admitting that invasive animal use raised a moral issue. "Animal use in research is not a moral issue, it is a scientific necessity," went a common dictum popular from the 1960s through the 1980s. With scientific ideology proscribing talk of subjective states in animals, common-sense acknowledgment of pain and other noxious states in animals was ruled out by fiat and was thus invisible, even to veterinary scientists. Ironically, ignoring pain and other mental states in animals led to bad science, as scientists disregarded the degree to which physiological, metabolic, reproductive, and immunological states in animals were affected by uncontrolled pain and distress, which had major physiological implications.

Though, by 1980, animal research was a major and controversial social issue, it had been defined in a way that admitted of no solution. The research community affirmed absolute entitlement to use animals as they saw fit; the opposition claimed that invasive animal research was tantamount to Nazi behavior. No middle ground was articulated.

By a series of fortuitous circumstances, my own career from the mid-1970s on has been linked to the issue of ensuring proper, morally based treatment of laboratory animals as a corollary of my philosophical interest in the moral status of animals in general and linked to the issue of how society would articulate its ever-increasing concern with animal treatment. I thus discuss this issue in a partially autobiographical way, as my colleagues and I were pivotal figures in helping society to express its concern for laboratory animals, while also getting the scientific community to acknowledge their moral status.

In essence we ended up articulating a middle ground in a legislative proposal. It was clear to me as a philosopher that the research claim to an absolute entitlement to use animals for human benefit was ill-founded philosophically, since one could find no difference between people and animals clearly morally relevant, in that animals were more like people than like wheelbarrows, in that what we did to them *mattered* to them – they are *sentient*. The lack of such a clear-cut difference rationally should give pause to scientists who wish to believe that we are unambiguously entitled to use animals, with all degrees of invasiveness, for human benefit. The researchers stance was also ill-founded in terms of social ethics, since society, as we shall shortly demonstrate, was expressing major concern over animal suffering.

At the same time it was clear that society also was unwilling to give up the health benefits it believed came via animal research. (This was clear because research was largely done with public money, and there was no social movement to rescind that funding.) Thus it dawned on me that neither extreme position (complete freedom in animal use vs. total abolition) was socially ethically viable. As I show below, we thus sought a middle ground.

In 1976, I was tapped by a number of socially prescient colleagues in the College of Veterinary Medicine at Colorado State University to develop a course in veterinary medical ethics, as these veterinarians had a strong intuition that society was changing rapidly with regard to developing social concern for animals. I was asked to articulate the elements of this change in moral terms for veterinary students, as well as to indicate where practices in the veterinary school curriculum were potentially at loggerheads with emerging ethics for animals.

Although courses in human medical ethics were well established by then, thinking about veterinary ethics and courses therein were nonexistent. What was called "veterinary ethics," like medical ethics a couple of decades earlier, or real estate ethics, or any other professional "ethics," was in fact largely issues of intraprofessional *etiquette* and not ethics at all. "Veterinary ethics," as then constituted, dealt with such trivial issues as fee-splitting, whether it was "ethical" to advertise, how big one's practice sign could be, and whether it was "ethical" to send Christmas cards to clients. The American Veterinary Medical Association's "Code of Ethics" contained dozens of references to advertising, and not one to the ethics of euthanizing a healthy animal for owner convenience, so-called convenience euthanasia. The veterinarian with whom I was partnered to teach the class, Dr. Harry Gorman, who was arguably one of the greatest veterinarians of the twentieth century, having been an experimental surgeon who invented the artificial hip in dogs, who commanded the aerospace program's use of animals, and who had been one of the highest ranking veterinarians in the military, was an intuitive philosopher who shared my interest in "real" veterinary ethics and my contempt for self-serving questions of etiquette. In fact, Dr. Gorman predicted correctly that discussions of advertising issues would soon be made moot by court decisions, which is exactly what occurred.

As I prepared to teach the class, my efforts focused on two related areas. In the first place, I needed to determine *whether* society was in fact growing increasingly concerned about animal treatment, an empirical question. And if society was indeed focusing morally on animal treatment, what form would that new ethic take? Second, I needed to examine the practices employed in veterinary education to see whether they would be considered acceptable to the new social ethic for animals, and endeavor to replace them with morally acceptable alternatives if they were not. This latter problem was a major one – how could I, a nonveterinarian "carpetbagger," a philosopher, an Easterner, make any headway with cowboy veterinarians on ethical matters in a western, redneck school?

By 1978, I had developed a set of answers to the above questions, published in 1981 as *Animal Rights and Human Morality*. In the first place, it was clear that in fact Western societies had begun to worry about animal treatment in an unprecedented way beginning in the mid-1960s, a trend that has increased exponentially in the ensuing years and continues today unabated and indeed strengthened.

Let us begin by looking at some milestones over the past thirty years evidencing the rise of a worldwide social concern for animal treatment. According to both the National Institutes of Health and the National Cattlemen's Association – hardly radical groups – who I consulted in 1993 while preparing a report explaining social concern for animals to the United States Department of Agriculture – from the late 1970s to the 1990s Congress received more letters, phone calls, and other communications on animal welfare–related issues than on any other social matters. In the same vein, in the early 1980s I was contacted by the Undersecretary of Defense for Health Affairs for help when the Defense Department announced that it would shoot a dozen anesthetized dogs in order to teach battlefield surgeons to deal with trauma inflicted by tumbling bullets. After this announcement, the department received more letters of protest than it had ever before received on any other matter, including one from the CEO of a major corporation that began "Shoot Secretaries of Defense, not defenseless dogs." The debate got progressively hotter.

In Europe, beginning with a sweeping 1988 Swedish law phasing out "factory farming" over fifteen years, the European Union has followed suit eliminating veal crates, sow stalls, and battery cages for laying hens. All Western European countries developed laws protecting laboratory animals.

In the United States, which tends to be naïve about agriculture, with most citizens thinking that farms are still Old McDonald's farms rather than factories, the confinement or "white" veal industry has nonetheless been virtually destroyed by consumer rejection of their practice of keeping calves near-anemic and confined to keep the meat white and tender. In November 2002 Florida voters passed a law banning sow stalls for pig production, though this was largely symbolic, since there are virtually no confinement swine operations in Florida. Beginning in the late 1990s, the activist group PETA forced McDonalds, Burger King, Wendy's, and other fast food restaurants to demand animal welfare improvements from their suppliers under pain of bad publicity and, as of this writing (2005), are also making similar demands of large supermarket chains.

Twenty-five years ago, one would have found virtually no U.S. federal, state, or local legislation pertaining to animal welfare with the exception of anticruelty laws. Today the situation is markedly different: Congress now sees some fifty to sixty animal welfare bills per year, ranging from protecting dolphins from tuna nets to preventing redundant experimentation in animal research. When I lectured at the Houston

Livestock Show in 1998, an official of the American Quarter Horse Association, the largest equine group in the country, informed the audience that the single largest expense the organization incurred was hiring a research firm to track various legislative proposals related to equine welfare, with the resultant volume being as thick as the Manhattan phone book. There were 2200 state laws proposed relating to animal welfare in 2004.

The Body Shop has become a multibillion-dollar industry by disavowing animal cosmetic testing. Circuses have waned in popularity because of perceived animal abuse, and the most popular circus in the United States is the Cirque du Soleil, which disavows animal acts. Cage-free eggs and natural meats are growth industries. BST (bovine somatotropin) to increase efficiency of milk production has been banned in Canada, Australia, New Zealand, and Europe in part because of the belief that it increases mastitis in cows. The British supermarket chain Tesco has adopted animal welfare guidelines far stricter than any British law mandates and maintains a team of veterinarian inspectors to audit producer compliance. Both the Vatican and U.S. Senator Robert Byrd issued strong statements against factory farming in the late 1990s. Markets such as Whole Foods that sell humane products are growing rapidly, as is the humanely oriented chain restaurant company Chipotle.

Wildlife management professionals find themselves ever-increasingly at loggerheads with society in general. Traditionally, wildlife managers managed game populations for hunters and fishermen, who paid their salaries by way of license fees. In the last fifteen years, "nonconsumptive" users of wildlife have far outnumbered hunters. Mountain lion hunting was abandoned in California, and steel-jaw traps were abolished by public referenda across the United States. In Colorado, the spring bear hunt was abolished by a constitutional amendment referendum when the politically appointed wildlife commission refused to heed recommendations from the Colorado Division of Wildlife to eliminate such hunts, where lactating mothers can be killed and the cubs die slowly of thirst and starvation. (The referendum passed 75% to 25%.) In Ontario, Canada, the Minister of the Environment abolished such bear hunts when surveys indicated that 90 percent of the public strongly opposed them. Wildlife managers have warned of "management by referendum" if they fail to accord with society's growing moral concern for animals. The British Parliament has passed a fox-hunting ban.

Zoos that are little better than prisons for animals – the state of the art in my youth – have been abolished by public concern. Greyhound and horse racing are waning, in significant measure for reasons of animal welfare. Bullfighting in Spain is under attack by the young. Hunting safaris in parts of Africa have been abolished in favor of photographic safaris. It has become politically incorrect to wear fur. Municipalities such as San Francisco and Boulder, Colorado, have declared that people with companion animals are not "owners," but "guardians." Malpractice awards against veterinarians have been increased beyond the market value of pets on the grounds of their emotional value to people.

Cruelty to animals has been elevated from a misdemeanor to a felony in thirty-nine states. Switzerland came close to banning biomedical research on animals, but such a ban was prevented at the last minute by drug companies' spending millions to change public attitudes. Sociologist George Gaskell has shown convincingly that European rejection of biotechnology is driven more by animal ethical concerns than by fear of catastrophe, the latter being the commonly held explanation. We return to this below in our discussion of biotechnology. California voters have made it a felony to ship a horse to slaughter or sell it to someone who will ship it for that purpose. Twenty-four law schools have courses in animal law, in part devoted to raising the status of animals from property to at least partial personhood. As of May 2002, the German Constitution was amended to extend constitutional protection to animals, particularly the right not to suffer at human hands unless it is absolutely essential. (The German biomedical community referred to the day that law was passed as "Black Friday.") The British Home Office has written a "Bill of Rights" for pets, the violation of which can result in forfeiture of the animal.

Surveys have indicated that well over 80 percent of the U.S. population believes that animals have rights, that an animal's life matters to it as much as a human's matters to us, and that public support for biomedical research declines in proportion to the pain suffered by the animal. As early as the 1970s, readers of *Glamour* magazine indicated that the animal suffering involved in safety testing was not justified by the development of new cosmetics. When a mountain lion killed a jogger in California and was subsequently shot, far more money was collected for the orphaned cubs than for the orphaned children!

The roots and signs of all this were evident to anyone who looked carefully for such evidence in the 1970s (though few bothered to look). Charged as I was with anticipating trends, I looked carefully and was thus cast in a prophetic mode. Biblical prophets such as Isaiah and Jeremiah were not crystal ball gazers or even people who foretold the future like Nostradamus. Rather, they were perceptive about current trends, and then could affirm, "If things continue as they are, such and such will be the inevitable outcome." In the same way, I saw a society ever-increasingly concerned about animal treatment for clear reasons we shall detail shortly.

In any event, I was convinced that attention to animal treatment would continue to develop in society. Furthermore, I was also convinced that the traditional social consensus ethic for animal treatment was totally inadequate to emerging concerns. (I had begun to study that ethic in preparation for teaching my veterinary school course.) Essentially, the consensus social ethic stretching back to biblical times and encoded in the legal system of all civilized countries since the early eighteenth century was an ethic of anticruelty – the prohibition of deliberate, willful, purposeful, deviant, intentional, sadistic infliction of unnecessary suffering on animals or outrageous neglect, such as failing to provide food or water. Historically, there were two moral reasons for this ethic: The rabbinical interpretation of biblical prohibitions against cruelty rested in the concept of *tsar baalay chaim* – the suffering of living things, which was recognized as an evil. The Catholic tradition, codified by St. Thomas Aquinas, was more oriented toward humans, in recognizing the psychological fact that those who perpetrate cruelty on animals will likely go on to become cruel to people, an insight confirmed by over twenty-five years of research linking childhood animal cruelty with adult psychopathy, child and spousal abuse, sadism, and even violent criminality and serial killing.

Until cruelty to animals was elevated to a felony in many states during the last two decades – and even subsequent to that elevation – cruelty was ignored or dealt with very leniently by overworked and overtaxed police and courts. For example, two of my veterinary students were prohibited by their lease from having pets. Despite the lease, they acquired a kitten. When the landlord found out, he let himself into the apartment, beat the kitten to death with a hammer, and wrote the students a note explaining his "solution to the problem." When they brought cruelty charges against him, he was convicted and fined $25. As he left the courtroom he leaned over to the students and said, "For $25, I'd do it again."

The biggest problem about the anticruelty ethic's being able to accommodate the new socio-ethical concerns of the last thirty years, however, is a conceptual one. If one makes a pie chart representing *all* the suffering that animals experience at human hands, and one then asks how much of that suffering is the result of deliberate cruelty, one will probably respond, as 100 percent of the audiences I have addressed on animal ethics have replied, "only a tiny fraction," "under 1 percent." Few of us have ever even witnessed deliberate cruelty. If we recall that the United States alone produces 14 billion chickens a year, and that some 80 percent experience fractures or bruises, one can readily see that cruelty is not the major issue about animal suffering. Most animal suffering results not from psychopathy but from perfectly acceptable motives: curing disease, learning about biology, producing cheap and plentiful food, and keeping consumers safe from toxicity. Agriculture, science, toxicology, hunting, trapping, and all socially accepted animal uses are exempt from the purview of the anticruelty laws. It was thus clear to me that if society continued to worry about animal treatment, it would need new ethical tools "beyond cruelty" to express its concerns and its remedies.

In our earlier discussion of ethics we explained the concept of ethical judo, what Plato called "recollection," as opposed to ethical sumo, or "teaching." The best way to win an ethical argument or to effect social change is to show one's opponent – or society – that the person or group already believes what you are pressing on them, but hasn't realized it. I had occasion to utilize this approach as I learned (to my horror) of some of the practices employed in veterinary schools to teach veterinary medicine. What was particularly morally outrageous to me was the then universally accepted practice among veterinary schools of teaching surgery by operating on the same animals over and over again, in what was called multiple survival surgery. In 1978, when I started teaching in the veterinary school, they were doing eight surgical procedures on the same animal, with virtually no provision for aftercare. CSU was a "good place," I was told – some schools did more than twenty surgeries. Many times these "procedures" included breaking jawbones and femurs to learn to repair them. The students were terrified by the stories they had heard and asked me to look into this issue.

As I rhetorically asked earlier in our discussion, how could I, an outsider, criticize these practices and get a hearing? The value of recollection – of ethical judo – was, by happy coincidence – revealed to me as the key to ethical change. One day, shortly after learning of the practice of multiple survived surgery, I ran into some veterinary surgeon colleagues in the gym. In an attempt to learn more about this practice, I guilelessly asked them, "Is this the only way to teach surgery?" I was amazed by the response this evoked. "Do you think we *like* to do this?" they roared. "Do you think I went $60,000 into debt to get a veterinary education just to treat an animal as something to whittle on?" "We hate it! It is simply a matter of money." Suddenly one of them had a thought: "You are the new ethics person. You fix it!" "I will try," I replied, "If you guys help me." When it became clear that the faculty considered the practice morally reprehensible, we were easily able to convert the teaching program to a single survival surgery with the students graded equally on "carpentry" and aftercare. Within two years, the surgeons had continued their ethical reflections to the point where they told me they could no longer justify *any* survival surgery and were willing to make the surgery labs terminal to avoid pain and suffering for the animals. "The students will learn recovery on clinical cases," they told me. The "CSU ethic" was soon adopted by most other veterinary schools. With this barrier crossed, it was relatively easy to eliminate other invasive and demoralizing laboratory exercises, some of which exist merely to "cull the sensitive," something veterinary medicine surely did not wish to do in the changing social world I described to them as ever-increasingly concerned with animal suffering.

Many years later, the *Denver Post* weekend magazine did a major story on my work with the CSU veterinary school, written by an excellent young reporter named Steve Singular, who was later to write a bestseller about the shooting of Denver radio talk show host Alan Berg by American neo-Nazis. Steve spent a few days with me, and talked to faculty members about my work. When we parted, he remarked that I had benefited a great deal from the "cowboy ethic" at CSU.

"You could never do what you do at Harvard," he remarked. "You have been fortunate to work at CSU which was dominated by a Western ethos." "What do you mean?" I asked. He laughed. "The cowboy ethic is to hear a guy out before you hang 'em! At Harvard, you would just be hanged for your lack of relevant credentials!" As a Columbia University product, I immediately saw his point. It is no surprise that much innovative applied ethics has come from land-grant schools with a strong agriculture tradition.

Once I was, as it were, reminded about recollection, I could deduce how society would articulate in moral terms its emerging concerns about animals. Ethical change, it was clear, would take place by recollection, with "new ethics" emerging from established ethics. If society wished to provide protection from animal's interests' being crushed by human convenience, it would clearly be sensible to apply the moral categories we use to protect individual *human* interests from being submerged under utilitarian pressures – the concept of *rights.* As we saw, rights are legally articulated moral notions that protect certain fundamental aspects of human nature from being overridden for the common good. For example, we may speak our minds even if we anger most other members of society and even if it costs society a large amount of resources to protect the unpopular orator – witness society allowing the American Nazis to speak in Skokie, Illinois, home of many Holocaust survivors. The Bill of Rights protects those aspects of a human essential to their humanity: not wanting to be tortured, holding on to one's property, believing as one chooses, and so on. Animal nature (or *telos,* as I have called it, following Aristotle) is clearer and easier to identify than human nature – the "pigness" of the pig, the "dogness" of the dog. Being with others of its own kind or free to forage is as important to some animals as speech or religion is to humans. So it seemed clear to me that if society wished to assure that animals used by humans lived decent lives, it would mandate legal protections for key aspects of animals' natures. As the proliferation of animal laws we mentioned earlier shows, that is precisely the direction society went. The *New York Times* called the Swedish law abolishing confinement agriculture a "Bill of Rights for farm animals."

Social concern in the United States was directed first toward the use of animals in science and medicine, research and training, as opposed to in Britain and Europe, where factory farming seemed to activate the greatest concern. There are a number of reasons for this. First, intellectuals have always been viewed with suspicion by Americans but have been greatly respected in Europe. (See Richard Hofstadter's excellent study *Anti-Intellectualism in American Life.*[1]) Adlai Stevenson is often considered to have lost his bid for the presidency because he was accurately but devastatingly depicted as an "egghead." Second, given the size of European countries and Britain, it was much harder to hide factory farms, as we have done in the United States, in remote areas. Third, the British were alerted relatively early to factory farming by Ruth Harrison's exposé, *Animal Machines*[2] and by a good deal of media coverage. (In the United States, agriculture is very poorly covered in the press.) Fourth, as the respective responses of Europe and the United States to genetically modified foods indicates, food is much more an object of cultural consciousness in Europe, where in many countries, people buy fresh food from markets. Europeans dine; we *fuel* – hence some of the execrable prepared foods we consume for convenience.

At any rate, by the late 1970s, concern about the treatment of animals in experimentation was well covered by the media. In fact, a case can be made that the first major demonstration expressing moral concern for animals was orchestrated by activist Henry Spira among New York City office workers in the spring of 1969, when they would assemble at the Museum of Natural History to eat lunch and protest manipulative experiments on feline sexuality. (Though these experiments were not painful, they did stir the public imagination by utilizing electrodes implanted in the cats' skulls.)

Armed with our knowledge of burgeoning social concern with animals, anticipation that this concern would inevitably turn to legal protections or rights for animals, and above all collectively possessing over sixty years of animal research experience, a group of veterinarians and myself at CSU knew that research animals were not getting close to the best possible treatment compatible with their use in research. In fact, some of the animal treatment *militated against* good research results by stressing or hurting the animals. Social primates were caged singly in tiny cages in totally impoverished environments, making them chronically stressed; beagles used to study mammary cancer were housed

[1] Hofstadter, *Anti-Intellectualism in American Life.*

[2] Harrison, *Animal Machines.*

in large kennels where the barking was so loud that caretakers and researchers had to wear ear protectors, though the dogs did not, and an article had appeared in *Science* showing that beagles subjected to unmitigated stressors such as noise developed more mammary tumors than unstressed animals.[3] When I did a literature search on analgesia for laboratory animals in the early 1980s, I could find *no* articles on the subject, save for one affirming that there ought to be such articles. When Dr. Harry Gorman took the job as experimental surgeon at CSU and went to the veterinary school pharmacy to get opiates to alleviate pain in the dogs he used for orthopedic experimentation, he was told that "we don't carry that" and was advised to "give them an aspirin." Veterinarians were totally untrained in pain management. Researchers often confuse paralytic drugs (i.e., ones that prevent movement while the patient remains conscious and able to feel pain) and anesthetics, as when one of my students protested against a state Fish and Wildlife agency doing caesarean sections on moose using succinylcholine chloride (a paralytic) and his concerns were brushed aside by scientists who affirmed that if the animals felt pain they would be thrashing around. True to scientific ideology, skepticism about animal awareness was rife and ignorance about ethics in animal research pervasive. (I discuss the effect of scientific ideology about pain in detail in a separate chapter.)

In one textbook of abnormal psychology, a photo of a rat was shown with the caption, "for ethical reasons, animals are used in research."[4] A student could get an M.D. or Ph.D. in an animal-using research area and learn nothing whatever about the animals he or she used except that they model a particular disease or syndrome. (As a result, I heard the same anecdote on three continents: researchers complaining to veterinarians that the latter had "supplied them with sick dogs, because all have temperatures over 98.6°F!")

Though all of the above problems represented difficult and serious ethical issues, as did indeed the general question of the moral justification for hurting innocent animals for human benefit, these issues were essentially never discussed intelligently in scientific circles, and indeed were never discussed at all in scientific journals, meetings, or classes. A personal anecdote illustrates the appalling lack of moral intelligence

[3] Riley, "Mouse Mammary Tumors."

[4] Rosenhan and Seligman, *Abnormal Psychology*, p. 154.

when scientists did occasionally attempt to engage in ethical defense of animal research.

I had been asked to speak to an AALAS meeting in Salt Lake City in 1980. The membership of the American Association for Laboratory Animal Science consists of laboratory animal veterinarians, animal researchers, Ph.D.s, M.D.s, technicians, and animal research industry people. They were naturally interested in legislation affecting their careers. Curiously, although laboratory animal veterinarians complained bitterly that they could improve the lot of research animals and the quality of science if researchers, particularly M.D.s and Ph.D.s, would listen to them, which they would not, these veterinarians nevertheless refused to support the very legislation that would empower them to improve things. The reason for this state of affairs was that they followed the lead of the human medical research community, whose attitude toward the issue of laboratory animal welfare was a hostile one, seeing the issue as a smokescreen for what they saw as antiscience crazies hell-bent on disrupting Western civilization. (One veterinary school dean actually expressed this attitude when some students asked for an alternative to killing a dog for their education.)

Not only was ethics ignored, but, as pointed out above, literature about and use of the analgesics and identification of pain in animals were nonexistent. Another AALAS anecdote displays the depth of the problem. In the early 1980s, I was invited to lecture at AALAS and to justify why I believed legislation to be necessary. There were five eminent laboratory animal veterinarians on our panel as well. In my presentation, I challenged them to identify an analgesic regimen for a rat used in a crush injury experiment. Their response was uniform and rooted in scientific ideology: "How can we do this if we can't even know the rat feels pain?" In my demand for laws forcing such control of pain, they affirmed, I was assuming what I needed to prove – namely, that animals felt pain. I return to this incident below.

Between the ethical ignorance of the medical community and the inflammatory rhetoric of animal activists, the social debate about research animals shed more heat than light. Standard debates were repeated with monotonous regularity, with activists claiming that animal research had never produced any benefits and researchers invoking Banting and Best's dog work leading to the discovery of insulin.

In fact, I attended one such debate in Australia in the early 1980s. The moderator, an intelligent and articulate law school dean, opened the conference by stating some ground rules. "We will have no claims that biomedical research or animals has produced no benefits, nor will we have lists of the benefits it produced. We will assume it has produced benefits and worry about the ethics of such invasive research." Her rules notwithstanding, the first speaker, one of the Australia's top biomedical researchers, began with Banting and Best and continued to list benefits.

On one occasion in the late 1970s, I debated the head of an extreme research defense group, the National Society for Medical Research. His opening statement was "Thirty-five Nobel Prize winners have used animals in research; therefore, it is permissible to use animals as we wish in research." The rest of his talk went downhill, as he turned to the audience of students. "You want laws to protect animals? Most of you will go home and smoke marijuana, breaking the laws we already have!" On another occasion, at about the same time, I was lecturing at a prominent veterinary school, and pointed out that by doing multiple surgery on animals, they were violating the National Institutes of Health Guide for the Care and Use of Laboratory Animals, which as recipients of federal funding, they were theoretically contractually obligated to, though NIH did not enforce it. The dean leapt up and said, "I wrote those guidelines." "Then presumably sir," I affirmed, "you ought to be exemplary in adhering to them." "Oh, I didn't mean it for people like me who know what they are doing," he said, "I intended them for some yokels in the Midwest."

These anecdotes illustrate the total inability of defenders of research to mount proper ethical arguments, given their unthinking adherence to their scientific ideology. Ethics was, in their view, a matter of emotion, not reason, so one tried to win on emotion. In fact, one of the research community's major rhetorical devices was the film mentioned earlier entitled *Will I Be All Right, Doctor?* a statement made by a frightened child to a pediatrician. "Yes, you will be all right," was the message, "provided that society leaves us alone to do what we want with animals." The film then purports to show the ideal treatment provided to research animals, including a sequence on surgery ironically showing many violations of proper aseptic technique. In fact, I saw the film premiered at a special session for laboratory animal veterinarians at an AALAS meeting. Though the people who sponsored the film wanted comments, there was only one. A laboratory animal veterinarian from Pennsylvania affirmed, "I believe that I speak for all of us. I am ashamed and appalled to be associated with a propaganda film pitched lower than the worst anti-vivisectionist clap-trap."

Though these anecdotes are bitterly amusing, they eloquently attest to the total inability of the scientific community to defend animal research in ethical terms. Scientific ideology ruled – one wins ethical arguments by shouting loudest or tugging at heartstrings. Almost all philosophers writing on the issues of animal research were uneasy for various reasons about our right to use animals invasively for our benefit. One exception was Michael A. Fox, a Canadian philosopher who attempted a defense of animal experimentation and the publication of whose book was financially assisted by the biomedical research community. Shortly after the book was published, Fox himself disavowed the arguments of the book and affirmed the immorality of invasive animal use. Another philosopher, Carl Cohen, won a visiting professorship at the Harvard Medical School after publishing an article in the *New England Journal of Medicine* arguing that instead of attempting to reduce the use of animals, we should use more of them (else we are likely to use humans).[5] Most philosophers felt that Cohen's arguments were weak or question-begging. For example, in his discussion of my work, he attacked me without acknowledging the replies I had already published in response to the sort of strategy he attempted.

Unlike most parties to the debate on animal research, my colleagues and I, in drafting the theory and practice of legislation, did not take an extreme position. We recognized that, in a democracy such as ours, sensitive to a full range of opinions, all social change would necessarily take place incrementally. This point had in fact already been recognized by animal activist Henry Spira. Thus our legislative approach was moderate, largely based on attempting to subvert researcher agnosticism about ethical and pain issues invisible to researchers because of their ideological blindness. We essentially believed that if we could get scientists to go beyond ideological denial of ethics and pain and suffering imposed by scientific ideology, they would be led to

[5] Carl Cohen, "The Case for the Use of Animals in Biomedical Research," *New England Journal of Medicine* 315, no. 14 (1986): 865–70.

"reappropriate common sense." Our strategy was one of enforced self-regulation, whereby we mandated local committees – animal care and use committees – to review prospectively all proposed animal research in ethical terms and to monitor the institutional animal care program, working in tandem with a laboratory animal veterinarian designated to supervise institutional animal care. The law was to give authority and "muscle" to the veterinarian, who historically had neither authority nor power to enforce proper husbandry, animal care, anesthesia and analgesia, and so on on researchers. The committee system, now mandated by federal law, gave veterinarians the requisite clout.

The thinking behind our law was meant to subvert scientific ideology, in a way that researchers, we hoped, would own, because they had created it. First of all, we did not want "a cop in every lab." We were familiar with the British system that relied on inspectors, who ended up more sympathetic to researchers than to the animals. Even more important, we wanted to help establish a generation of ethically conscious researchers, who thought automatically in moral terms and in terms of animal pain and who would help their students to achieve the same degree of awareness. Our strategy was first of all to mandate local review committees consisting of scientists, nonscientists from universities or the research establishment, and lay people. They would prospectively review protocols and inspect facilities. The review would look at whether the experiment was well designed, whether it utilized the correct species and numbers of animals (neither too many nor too few), whether alternatives to animals had been considered, and, most important, whether pain and suffering and distress were properly controlled in invasive work and postsurgically by judicious use of tranquilizers, sedatives, anxiolytics, local and general anesthetics, and, most revolutionarily, analgesics. We specified these requirements in very broad terms and left it to the scientists to work out precise applications. It was our belief that such an approach would force ethical thinking and proliferate it. Rather than handing down ethical chapter and verse, which could be rejected the way human researchers ignored detailed social ethics for human subjects, we felt that asking the scientists to work out individual applications would give them a sense of ownership in what they created, the results of which would also be spread at scientific meetings, in journals, and later via the Internet. We replaced preformed rules with ethical dialogue, which we hoped would trump scientific ideology and force the reappropriation of common sense. In our version of the law, we demanded accommodations for animals which fit their biological and psychological natures; only an estimated 10 percent of research protocols are invasive or painful, but 100 percent historically involved keeping animals in impoverished environments.

Congress did not heed our advice, but instead required only exercise for dogs and living environments for primates that "enhance their psychological well-being." Later, however, NIH was to rewrite its *Guide for the Care and Use of Laboratory Animals*,[6] the "bible" for animal care and use committees, to place great emphasis on enriching environments for all species. In serving on ACUCs, and in filling out protocol forms, scientists were forced to begin thinking about both ethics and animals' subjective experience, to accord better with social concerns, and to reappropriate common sense.

In 1980, Colorado State University voluntarily adopted the regimen we proposed for our laws, so that we could gather experience with its use to present to Congress. And in 1982, in the face of highly publicized abuses of laboratory animals at a private laboratory in Maryland and other revelations of atrocities perpetrated on laboratory animals, I was asked to testify before Congress on a week's notice on behalf of our proposed law. (In 1984, videotapes made by researchers at the University of Pennsylvania of head-injury research done on baboons were stolen and released to the media by activists, further galvanizing public support for legislation.) Utilizing connections to the research community, I prevailed upon the American Physiological Society, the traditional opponent of any intrusion into the research process, to support our legislation, and also received support from CSU, UCLA, and the University of Florida. At the time, there was another piece of legislation being supported strongly, entitled the Research Modernization Act, which would have cut the biomedical research budget by up to 60 percent and plowed that money into alternatives to animals. Unfortunately, that bill was very naïve scientifically, as by "alternatives" the bill's chief architect meant "building an artificial dog that howls when cut and bleeds ketchup." She had further suggested that if we

[6] Institute of Laboratory Animal Resources, *Guide for the Case and Use of Laboratory Animals*.

could send a man to the moon, we could "surely" model a mouse on a computer. She, of course, failed to realize that the latter was far more difficult than the former – if we know enough about mice to model them on a computer, we probably would not need to. Fear of such scientifically naïve – and destructive – mandates made our approach to "enforced self-regulation" more attractive to reflective scientists.

It was not, however, attractive to everyone. We were still largely opposed by the wealthy and powerful medical research community. When I was invited to Congress to appear before Representative Henry Waxman's committee, I was told that the immensely popular Dr. Donald Kennedy, former FDA commissioner and then president of Stanford University, would speak in opposition to the bill. Calling back to contacts at Stanford, I found that they had an awful record of violations under the current Animal Welfare Act, a listing of which I quickly received from the USDA under the Freedom of Information Act. I then sent word to Stanford that if Kennedy chose to oppose our bill, I would read the Stanford violations into the Congressional Record. This seemed to lead Kennedy to reconsider. He was, however, replaced by pioneering heart surgeon and president of Baylor, Dr. Michael De Bakey. Out of the frying pan, I thought...

The day of the hearings arrived and De Bakey spoke first. I was amazed to find that he had read neither the bill nor the supporting testimonials I had garnered. This is, he said, a situation of well-meaning but scientifically naïve people mixing in matters they don't understand. The authors and supporters, he continued, were totally unaware of how this would devastate research and human health. And he continued in this vein. I had the pleasure of responding and pointing out that the naïve supporters he cited included major research universities and the American Physiological Society, the traditional opponent of any intrusion into research.

This was not the first time we had encountered silly responses. As mentioned earlier, we originally naïvely wrote the bill for the state of Colorado, not realizing that such laws needed to be federal, else researchers would go to other states. In any case, we were killed by the Agriculture Committee of the Colorado legislature, on the dubious grounds that "today it is lab animals and next you'll be telling us how often we can use a hot shot on our cattle." Following that disappointment, we had been contacted by Colorado Representative

Pat Schroeder, who offered to carry the bill federally. Although by her own admission she was lukewarm on the project, her enthusiasm increased when she received a bizarre letter from a medical school dean. This bill is unneeded, intoned the dean in his first paragraph, because everything it stipulates is already done at our medical school. And if it passes, he continued in the second paragraph, all research in our school would have to stop. "It was then," said Representative Schroeder, "that I smelled a rat."

Sure enough, De Bakey later told me and Representative Doug Walgren, who was sponsoring the bill in 1982, that he had been told merely that this was more radical antivivisectionist activity and to give your "research is important" speech. Subsequent witnesses from Congress included Senator Bob Dole and Senator John Melcher, then the only veterinarian in Congress. They and virtually all other witnesses enthusiastically supported the bill, and I was told by Henry Waxman and other committee members that although it would not pass this time, its eventual passage was assured by public opinion. Sure enough, two pieces of legislation based on our model passed in 1985, one amending the Animal Welfare Act and the other requiring NIH to follow the same principles for federal research funding recipients. The NIH provisions allowed seizure of all federal money from an institution violating the law, and suddenly the law was serious business as NIH made an example of delinquent institutions. The final versions of the law would of course require the writing of detailed regulations by USDA to interpret it, but at least project review, ACUC inspections, and, above all, the mandate to control pain were off and running. The laws went into effect in January of 1986.

Particularly perplexing to a scientific community held in thrall by scientific ideology were the requirements to control pain and to provide living accommodations for primates that "enhance their psychological well-being." The head of the Animal and Plant Health Inspection Services (the branch of USDA changed with writing detailed regulations interpreting the law, as well as enforcing it), Dr. Robert Rissler, freely admitted that, as a veterinarian, he knew nothing of primate psychological well-being, so he approached the primatology section of the American Psychological Association, all of whom were agnostic about animal consciousness. "Don't worry," they told him, "there is no such thing!" "Oh but there will be after January 1, 1987, whether you people help me or not!" rejoined Rissler. In the same vein, the research community complained to Congress that it had no clue as to how to identify or control pain in animals. Congress then approached the American Veterinary Medical Association (AVMA): "You are veterinarians. You know about animal pain. Put on a conference and issue a report for the research community." Despite the fact that most veterinarians were in fact agnostic about animal pain and knew little, the AVMA was forced to agree. (In 1982, the American Physiological Society had, in the same vein, held a conference and created a book entitled *Animal Pain: Perception and Alleviation* to reassure the public. In fact, virtually all of the book dealt with the "plumbing of pain," and only one small part of one paper dealt with the fact that pain hurts.)[7]

I was asked to speak at the AVMA pain conference and serve on the panel that issued the report.[8] It was a classic exercise in abandoning scientific ideology and reappropriating ordinary common sense. The chairman of the panel was Dean Hiram Kitchen of the University of Tennessee veterinary school, a sophisticated, philosophical, and humane man, but also a tough overseer. "Rollin," he growled, "you're a philosopher and a bullshitter, so you go write the introduction." I complied and stressed the degree to which ordinary common sense takes it for granted that animals have mental lives, and I pointed out that even the greatest philosophical skeptic of all times, David Hume, who questioned the reality of mind, body, causation, scientific knowledge, and religion, took it for granted that animals were aware. I read this to the group and was rewarded by a scientist from the National Institutes of Mental Health standing up and saying, "If we are going to talk that kind of nonsense, I am leaving!" Another member of the group, however, began to muse out loud. "Rollin may have a point," he said. "If animals are pain models for humans, then humans are also pain models for animals. So maybe it's okay to be somewhat anthropomorphic." "Yes," I shouted, "brilliant point!" – though part of me was cursing under my breath: "Yeah sure, I wrote the same point fifteen years ago and it went over like a lead balloon." Nonetheless, I continued to

[7] Kitchell and Erickson, *Animal Pain.*

[8] Kitchen, Aronson, Bittle et al., "Panel Report on the Colloquium on Recognition and Alleviation of Animal Pain and Distress."

encourage such "reappropriation of common sense," and the final document that emerged was perfectly suited to lead scientists to acknowledge felt pain in animals.

There has been a considerable degree of "reappropration of common sense" with regard to animal pain since the laws passed. From no papers on animal analgesia in 1980, we have gone to literally thousands. Ironically, the latest edition of *Veterinary Anesthesia*, by Lumb and Jones, the first U.S. textbook on this topic, which in all its earlier editions did not even give control of felt pain as a reason for anesthesia, now has eloquent discussions of felt pain as a stressor and a strong emphasis on analgesia.

This revolution is again best illustrated with an anecdote. In the early 1980s, as mentioned, I was invited to present my reasons for wanting legislation at an annual AALAS meeting in Washington, D.C., on a panel with a number of prominent laboratory animal veterinarians. In my presentation, I raised the case of a rat being used in a limb-crushing experiment (such injuries are very painful). I challenged the veterinarians on the panel to tell me what analgesic they would use to control the pain. "How do we know," they said in essence, "we don't even know that the rat feels pain." "That is why we need laws," I replied. "You are begging the question," they said. "You can't say we need laws until you have demonstrated that rats feel pain." My appeal to common sense left them unmoved.

Half a dozen years later, after the laws passed and went into effect, I called one of the veterinarians who had been on the panel. "Well," I said, "you were agnostic about animal pain at the AALAS symposium. But now you *must* use analgesia. What would you use after the crush experiment on the rat?" To my surprise, he rattled off a series of analgesic modalities. Surprised, I said, "How do you know that now?" "Easy," he said, "We simply went to the drug companies!" "What do you mean?" I said. "Drug companies don't develop analgesia for rats." "No," he said. "But all human analgesics for years have been tested on rats. I just retrieved the data!" Although he knew this in the early 1980s, it was only the passage of the laws that caused him to shift his gestalt and to see that data as relevant to rat pain.

Not all scientists reappropriated common sense as easily as that veterinarian. Many need more "lubrication." As a result, I was approached by CRC Press to do a book providing the knowledge that scientists would need to follow the law. At first I demurred, telling the editor that the task was best done by some laboratory animal veterinarians and naming those I thought to be most suitable. "They suggested you!" said the editor. As a result, my career took an unexpected twist that has subsequently occurred repeatedly, as I learned that to be effective, an applied philosopher cannot merely criticize; he or she must help to provide alternatives to what is being criticized. So, with the help of nearly fifty colleagues in science and veterinary medicine, a laboratory animal veterinarian and I assembled a two-volume book of scientific chapters aimed at helping scientists to obey the ethics and animal subjective experience aspect of the laws.[9] The authors did a splendid job, and the book is still valuable. There are six chapters on pain and one on boredom, and most strike a note melding science and ethics. (The chapter on analgesia, for example, begins with a statement that it is now clear that animals have rights and that the most fundamental right is the right not to suffer pain while being used for research.) Countless scientists were helped to shift their gestalt by these conscientious authors.

One anesthesiologist colleague told me a similar story. After the laws passed, he was asked to lecture on recognizing animal pain in laboratory animals at a medical school. Precisely because he had never abandoned ordinary common sense about animal pain, he had great trouble writing this paper, which seemed "too obvious" to him. Finally, while on his way to the school, literally approaching the airport, he grabbed an envelope and jotted down a list of criteria such as "the animal is tender at the point of injury," "guards the limb," and so on. Greatly embarrassed, he made his presentation and was amazed to find that it was the most requested copy of a paper he ever produced.

In brief summary form, the laws contained the following provisions.

1. Establishment of an institutional animal care committee to monitor animal care and inspect facilities. Members must include a veterinarian and a person not affiliated with the research facility to represent public concerns.
2. Standards for exercise of dogs are to be promulgated by the USDA.

[9] Rollin and Kesel, *The Experimental Animal in Biomedical Research.*

3. Standards for a physical environment that promotes "the psychological well-being of primates" are to be promulgated.
4. Standards for adequate veterinary care to alleviate "pain and distress," including use of anesthetics, analgesics, and tranquilizers, are to be promulgated.
5. No paralytic drugs are to be used without anesthetics.
6. Alternatives to painful or "distressful" procedures must be considered by the investigator.
7. Multiple surgery is prohibited except in cases of "scientific necessity."
8. The animal care committee must inspect all facilities semiannually, review practices involving pain, review the conditions of animals, and file an inspection report detailing violations and deficiencies. Minority reports must also be filed.
9. The USDA is directed to establish an information service at the National Agricultural Library, which provides information aimed at eliminating duplication of animal experiments and at reducing or replacing animal use, minimizing animal pain and suffering, and aiding in training of animal users.
10. The research facility must provide training for all animal users and caretakers on humane practice and experimentation, research methods that limit pain, use of the information service of the National Agricultural Library, and methods of reporting deficiencies in animal care and treatment.
11. A significant penalty is established for any animal care committee member who reveals trade secrets.
12. The USDA is directed to consult with the Department of Health and Human Services (which has responsibility for funding biomedical research through NIH) in establishing the standards described. Federal funding may be withdrawn from institutions that violate the law.
13. New civil penalties are provided for violation of the act.

In the end, do these laws work? The answer, I believe, is an emphatic "yes." Certainly they have led to the reappropriation of common sense about animal pain. From the absence of any knowledge of animal pain and the total failure to utilize analgesics we have gone to literally thousands of publications on animal pain and liberal use of analgesia, though we are by no means perfect. Equally important, as academic veterinarians and researchers utilize more pain killers, knowledge of and concern for animal pain increases in the next generation of researchers (i.e., Ph.D. candidates) and in private veterinary practitioners. Almost twenty years after the laws went into effect and the treatment and management and recognition of pain are well acknowledged in research, USDA has begun to look at "distress" – they wisely waited until scientific ideology had been at least somewhat eroded. The same is true of ethical thinking about animal research. When I now teach courses in science and ethics for Ph.D. candidates, far fewer are fully possessed by scientific ideology, and for those who are, it is far easier to dislodge. The filling out of protocol forms has greatly increased the discussion of proper number and species of animals, proper technique, pain control, and so on, and even if the research warrants pain and suffering, it is now widely known that pain is a stressor and can skew research results. Animal care and use committees have extended the moral logic of the laws to animal uses not mandated, for example, farm animals used in agricultural research, and invertebrates where we have reason to believe that pain might be present (note the emergence of a "give the animals the benefit of the doubt" mind-set). Although the laws do not mandate cost-benefit analyses on research comparing animal suffering to putative value of the research, such comparison has been done explicitly or implicitly at most research institutions from the very beginning, according to my sources at NIH.

One illustrative anecdote sticks in my mind very prominently. In 1998, I was invited to speak to the American College of Laboratory Animal Medicine, the board certifying laboratory animal veterinarians. As part of my duties, I was asked to debate publicly with a senior M.D./Ph.D. researcher of some prominence. I was arguing that the laws work; he was arguing that they do not. "Why do you say that?" I asked. "Well," he said, "I proposed a piece of research that involved creating a mouse model for a human genetic disease that is very painful and afflicts children. The committee in my own institution turned *me* down because I had no way to control the mouse pain." My reply was obvious: "It sounds to me like the system is working perfectly," I replied. The only evidence this scientist could muster against this claim was his own inconvenience, showing at most that at least one researcher had not been led to reappropriate common sense. (Some years later, I heard that the same researcher has experienced a gestalt shift and is now a major advocate for these laws.)

I do not claim that things worked smoothly from the outset. When my own institution's committee was formed in 1980, we would cover thirty protocols and eat lunch in an hour, with everyone loath to criticize a peer. I fretted a great deal. Had we just created a bureaucratic rubber stamp? Eventually, however, things changed by virtue of what I have elsewhere called "breakthrough experiences."[10] An example of such experiences occurred very early in the history of the committee. We were discussing a project involving the steel-jaw trap for use on coyotes, and everyone was treading very lightly. One of our members, a prominent and assertive veterinary surgeon, came in late. "Are we on that trap project? Well let me just say this! As long as this is a goddamn animal care committee, there is no way on God's green earth that I will *ever* approve a steel-jaw trap experiment – it is barbaric!" Everyone was stunned, and finally a wildlife biology researcher said, "I don't think it is any more barbaric than a later protocol where veterinary surgeons deliberately create cartilage defects to study lameness!" After then, we pulled far fewer punches.

A second example involves an animal scientist on the committee, who plainly saw the committee as nonsense, pandering to public opinion, and who never said anything. He was, however, shocked out of his dogmatic slumber when a protocol was submitted to drown pigs and study the time it took them to refloat, so insurance companies would know when to expect drowning victims to surface! He exploded: "What nonsense! This is a simple project in biophysics. We don't need pigs! All we need is intestine from slaughterhouse in which we can study gas formation!"

Finally, our first committee chair made it clear that he believed that all researchers, especially those who are veterinarians, knew what they were doing and did not need oversight. He was forced to change this view when we received a protocol where the euthanasia procedure involved using magnesium sulfate on cows, a method long forbidden by the American Veterinary Medical Association because it kills by paralyzing respiratory muscles. The researcher was a senior professor on the verge of retirement who had been doing research for well over

[10] Rollin, "Ethics, Animal Welfare, and ACUCs."

thirty years and should have known all about magnesium sulfate. To the credit of the chairman, he never again assumed that anyone knew what he or she was doing unless it was documented.

These breakthrough experiences work by changing people's gestalt, and thereby allow researchers to crack well-situated ideological carapaces. Suddenly, one realizes that not all science is intelligent, not all protocols are well designed, and, frankly, not everyone knows what he or she is doing. It quickly becomes clear that the *majority* of researchers don't know what they are doing when they try to justify the number of animals they use or when they don't know anything about the range of possible analgesic regiments, though they have been writing protocols and doing surgery for years. In actual fact, as alluded to earlier, one of the biggest problems creating animal suffering is that one could get an M.D./Ph.D degree from Harvard in biomedicine and never learn anything about the animals they use save that they model a particular disease or syndrome. Hence the horror stories one hears from laboratory animal veterinarians all over the world about researchers complaining of sick dogs because "they have temperatures over 98.6°F" or other researchers complaining that their guinea pigs have a "wasting disease" when in fact it is merely the case that they cannot chew (their teeth are maloccluded, or don't meet properly, and need to be trimmed) and thus do not receive the requisite nourishment.

Insofar as the law has forced centralization of animal facilities and, much more important, the oversight of research and animal care by trained laboratory animal veterinarians with ACUC backing and highly professional technical staff, things have improved greatly. In the old days, animal care was often done by student hourly workers, who would go off for a weekend and fail to feed or water the animals. And whereas, historically, laboratory animal veterinarians who pushed well-funded powerful researchers to do the right thing were often forced out of their jobs, this is much mitigated when the veterinarian is backed by federal law and a powerful ACUC made up of the best and most powerful researchers on a campus. Few researchers wish to risk the wrath of their peers or the federal funding to their institution.

Which brings us to a very common animal activist objection to these laws. "They mandate the fox to guard the chickens." Researchers are policing researchers, so how can the system work? For those tempted by this argument, I have a variety of responses. First, let us think back to junior high or high school. A smart teacher would deal with a problem student not by fighting him but by making him a monitor. When people serve on committees, they quickly lose their "researcher hat" and wear a "guardian of the public trust hat," which people invariably take seriously. At CSU, early in the committee's history, we received a scurrilous memo from a department head that has become known as "the infamous 'what is an animal' memo." This professor began by claiming that the committee cared only about "cute and cuddly" animals and was furthermore driven by sentimentality and emotion and catered to the vagaries of public opinion. And that was just the beginning. The committee responded by appointing this man to the committee. Within two years he became chairman, for ten years. When he left the committee we presented him with his memo, suitably framed. Within three years, he was back on the committee, and has served for about twenty years.

A few months ago, the committee experienced a different sort of breakthrough experience. A new investigator arrived on campus and submitted a surgical protocol for mice without providing an analgesic regimen. The committee rejected the protocol and demanded analgesia. "You don't understand," she complained. "Analgesia *might* affect my research." "Give us references showing that all possible regimens *will* skew your results," we demanded. "You still don't understand," she replied. "I don't *care* about the animals. They are just means to an end. Tools. I don't care about their pain."

She insisted on coming before the committee and telling us all this face to face. "What do you bunch of veterinarians know about real science?" she began. (In fact, only three out of twelve members are veterinarians, and two of those are Ph.D.s as well.) "I came from the famous Flotsam laboratories, and *we* didn't require analgesia." "Really?" we said, and immediately made it a policy for the campus not to allow the purchase of animals who had surgery at Flotsam. She continued to, as it were, dig her own grave. We were very nice to her, primarily because we were too shocked to be mean. When she left, two longtime members of the committee turned to me. "My God," they said. "Is that what we sounded like twenty years ago?" I nodded. "Well thank God we outgrew that," they intoned. "Thanks for helping us grow." Unless this researcher grows quickly, we suspect her days at CSU are numbered.

Around the country some committees have been able to get a small budget to fund research that benefits animal welfare. Early in our history, we faced a researcher who injected fractionated venom into mice to determine toxicity and extent of injury. He insisted that it did not hurt; we quickly determined otherwise. We then asked him to use anesthesia. "No way," he said. "It will distort my results!" "All anesthesia?" we queried. "I don't know," he said. "I've only tried a few." The committee funded him to try numerous other regimens, and he found that some did not affect his results. He published that data, and thereby changed the way such assays were done in his field.

The committees are not paper tigers, either. Constant clashes with committees on improper animal use can lead to loss of jobs or denial of tenure. In one case, in the face of outrageous and egregious behavior by a tenured full professor researcher, the committee imposed such draconian rules on him that he resigned from the university in six months.

Most telling, perhaps, is the extraordinarily long terms that people around the country serve on committees. Early in our history I realized we needed an expert in wildlife biology. I asked one of our best researchers, a prominent anarchist/cowboy, and a good friend, to serve. "I hate committees," he said. "They are a waste of time." "Not this one," I replied. "Give it a shot." He did so, and stayed on almost twenty years, until he retired. At his retirement party, he acknowledged that, by any measure, the ACUC was not a waste of time.

I do not mean to suggest that the laws are flawless and that all animal research issues are solved. At best, a mechanism now exists whereby researchers can (and hopefully must) engage such issues, and a crack has been made in scientific ideology.

In the course of the discussions engendered on committees, many ethical questions traditionally invisible to scientists have emerged: Is it indeed morally superior to do invasive research on "lower" animals than "higher" animals? How far can we attribute anthropomorphic morally relevant mental states to animals? If animals do experience pain, fear, boredom, anxiety, and other morally relevant states analogous to what humans experience, what right do we have to inflict them on animals when we would not do so to humans? And if the states are not analogous, why study them? (As of this writing, the USDA has begun to look at animal distress, as mandated by the law. They were wise to wait until pain recognition had become well established.) Is there an ethical issue about killing animals painlessly in research or only about hurting them? Is it morally correct for the Animal Welfare Act to be interpreted as affirming that rats, mice, and birds are legally not animals? Are there morally relevant differences between purpose-bred dogs and cats and random source animals used in research? (As an interesting exercise, the reader might try to construct other ethical questions about animal research that need to be answered.)

Biotechnology raises a number of issues that will challenge the ethics of animal research. For one thing, Congress has refused to extend the Animal Welfare Act to cover rats and mice and birds, though such animals are covered by the NIH law in research institutions getting federal money. The problem is that start-up biotechnology companies, sometimes doing very invasive things to rats and mice, are not covered at all.

Second, the laws allow research where pain cannot be controlled, such as pain research, though such studies usually bow to public opinion and try to restrict pain. Indeed, as early as 1980, the society for the Study of Pain formulated guidelines to limit animal pain in research.[11] More serious is an issue I shall discuss in another chapter: our ability to in principle genetically engineer animal models for human disease with horrific symptoms. Yet another problem is inherent in the nature of some genetic research. Such research ablates or adds genes to see what happens. What happens is almost never predictable; thus it is difficult to write prospective protocols for committees to look at. Committees must then demand early end points for animals that suffer.

Other problems arise from the fact that the laws do not cover "food and fibre" agricultural research, though many committees review such protocols without clear authority. Toxicology, too, raises problems, as international agreements regarding drug safety demand lethal or harmful dosing of animals for long periods of time. Such toxicological standards, believed by many toxicologists to be arbitrary and bureaucratic, should fold under public pressure.

[11] Committee for Research and Ethical Issues of the International Association for the Study of Pain, "Ethical Guidelines for Investigations of Experimental Pain in Conscious Animals."

In the end, though, animal research issues are at least being engaged, and the system seems to be respected. As we said earlier, this is counterintuitive in a scientific community that seems to fail to see human research as morally problematic. Can we learn anything from animal research about ethical self-regulation, which we can extrapolate to human subjects? I believe one lesson is that more draconian regulation does not assure moral thought; probably the opposite is true. The ACUCs work by a general mandate to think through ethical issues, with the details to be generated within science by dialogue and reflection as people deal with real issues. What people work to create, they feel ownership in. Perhaps we need to start afresh in human research, with a general mandate that in essence forces thought and reflection, not hoop jumping. Imposing the whole cloth of social ethics in terms of regulations on a scientific community trained to see it as irrelevant has not worked. Perhaps we need to create a system requiring dialogue and reflection about general moral questions, so that the research community can "recollect" human ethics by reinventing it and can be led to own what they create.

6

Biotechnology and Ethics I

Is Genetic Engineering Intrinsically Wrong?

If one were asked to name the area where the scientific community's ignoring of ethics has wrought the greatest harm to public acceptance of science, one would have to choose biotechnology. Genetic engineering, cloning, and stem cell research have all engendered equivocal public reactions at best, and outright rejection at worst, by virtue of scientists' consistent failure to articulate and engage the multitude of ethical and social issues the public believes to be inherent in the new modalities.

Any major new technology will create a lacuna in social and ethical thought in direct proportion to its novelty. What effects will this technology have on our lives? Will benefits outweigh costs, harms outweigh goods? What are the possibilities of the technology leading to more evil in the world, or in human society? How likely is it to be misused? Is it inherently wrong for any reason? Will it promote justice or injustice? Is it something benign or beneficial, something to be curtailed or allowed to soar? Will things be better or worse in virtue of its existence? Such questions inevitably bubble up in the social mind, and consequentialist questions are mixed up with deontological ones, issues of bad effects likely or possible to ensue are confused with inherent wrongness. Add to this unsavory mixture a liberal portion of religiosity, and one has truly created an intellectual Golem: powerful, mindless, and unstoppable.

I have called this situation a Gresham's law for ethics. Gresham's law, articulated in the sixteenth century by Thomas Gresham, affirms that "bad money drives good money out of circulation." If one has an economic situation such as the one prevailing in Germany after World War I, where it took a wheelbarrow full of deutsche marks to buy a loaf of bread, rational people will not pay their debts with gold. They will hoard the valuable currency and pay with the devalued currency worth next to nothing. Similarly, in the absence of anyone providing sound and reasonable articulation of possible ethical issues occasioned by a new technology, bad ethical formulations articulated by people with vested interest will rush to fill the vacuum created in the social mind by the new technology. Given that the scientific community is prohibited by scientific ideology from thinking through ethical questions in a rational way, the issues are defined, articulated, and decided by those who have an axe to grind: doomsayers, preachers, and Luddites, who set forth "ethical" claims that crowd out real ethical issues.

For example, when the cloning of Dolly the sheep was announced in 1997, Dolly's "creators" were silent on the ethics of cloning animals. As a result, within one week Time-Warner announced the results of a poll that indicated that 75 percent of the U.S. public believed that cloning the sheep had "violated God's will."[1] How, we may ask, could cloning advocates respond to *that* assumption or refute such a claim? This was compounded when one of the scientists involved in the cloning, having earlier disavowed any knowledge of the ethical issues that cloning occasioned, nonetheless averred that it would be clearly wrong to clone humans. There was no explanation, no argument. Shortly after that, doomsayer Jeremy Rifkin, who has made a career of opposing biotechnology, announced a pending coalition with fundamentalist Southern Baptists, mirabile dictu, to block cloning. President Clinton's Bioethics Commission, heavily laden with religious leaders, not surprisingly condemned cloning, particularly of humans.[2]

It seems eminently sensible that those who introduce new technologies should lead a dialogue on the genuine ethical issues they occasion. They may be wrong in how they articulate the issues, but at least they ought to be able to keep the discussion on a rational and informed plane. But alas, they usually fail to do even this. Consider the advent of computers. Norbert Wiener, a pivotal figure in that revolution,

[1] CNN/*Time* poll, "Most Americans Say Cloning Is Wrong."
[2] National Bioethics Advisory Commission, *Cloning Human Beings*.

saw few issues in the dawning of the computer age, suggesting in a book only that computers would usher in "the human use of human beings."[3] Few worried about the loss of privacy, the roaming of child molesters on the Internet, the vulnerability of banks to hackers, the loss of reading skills by a generation of children spoiled by computers. Perhaps some of these concerns were not predictable even if good minds had engaged in earnest discussion, but very likely others were. What ensued, however, filling the gap in social thought, was much concern about computers taking over the world" or "going mad," as in the film *2001* or the novel and film *Colossus*,[4] which helped society not at all to deal with the computer age.

The odd fact about Dolly was that I did receive a call from Dolly's creators some years before her birth was announced, asking me to discuss possible socio-ethical reactions to a cloned animal. I stressed that the public would first of all need to know whether cloning harmed the animal or not. Second, and equally important, the public would need fully to understand the technology well in advance of the announcement of such an achievement to prevent the Gresham effect. I suggested anticipatory stories on the reasons for, possibilities of, and possible concerns about cloning animals, in which as much as possible was explained without compromising proprietary considerations. My advice was effective only in the breach – Dolly was announced without any preparation at all, and we have already mentioned the outcome. We return below to an in-depth examination of cloning.

The need for the scientific community to educate the public on scientific advances and to lead the public discussion of ethical issues is well illustrated beyond the above anecdote. In particular it is made plain by the work of British sociologist George Gaskell at the London School of Economics, which we now discuss.

In June 1997, a team of researchers working as part of a Concerted Action of the European Commission and coordinated by George Gaskell of the London School of Economics, released the results of a survey of public attitudes toward biotechnology conducted in each of

3 Norbert Wiener, *The Human Use of Human Beings: Cybernetics and Society* (Garden City, N.Y.: Doubleday, 1956).

4 D. F. Jones, *Colossus Trilogy: A Twenty-fifth Anniversary Omnibus Edition* (New York: Image Publishing, 1992).

sixteen European Union countries.[5] According to Gaskell (personal communication), the results astonished the researchers, shattering both their preconceptions and conventional scientific wisdom about social responses toward biotechnology. The researchers found that "few [people] approve of the use of transgenic animals for research."[6] In addition, "there is a striking mismatch between the traditional concern of regulators with issues of risk and safety, and that of the public, which centers on questions of *moral acceptability*" (emphasis is mine).[7] Although conventional wisdom suggests that the overwhelming social concern about biotechnology is risk, the survey confuted that presupposition. When the 17,000 people surveyed were asked about six different aspects of biotechnology – genetic testing (using genetic tests to detect heritable diseases); medicine production, using human genes in bacteria to produce medicines or vaccines, as has been done with insulin; crop plant modification, for example, moving genes from plant species into crops to produce resistance to insects; food production, for example, to make foods higher in protein or have longer storage life; transgenic research animals genetically modified for research, such as the onco-mouse; and xenotransplants, introducing human genes into animals to render their organs immunocompatible for human transplants – all were perceived as potentially useful, but the uses of transgenic animals for research and transplantation were seen as morally unacceptable.

> The pattern of results across the six applications suggests that perceptions of usefulness, riskiness, and moral acceptability could be combined to shape overall support [for biotechnology] in the following way. First, usefulness is a precondition of support; second, people seem prepared to accept some risk as long as there is a perception of usefulness and no moral concern; but third, and crucially, moral doubts act as a veto irrespective of people's views on use and risk. The finding that risk is less significant than moral acceptability in shaping public perceptions of biotechnology holds true in each EU country and across all six specific applications. . . . This has important implications for policy making. In general, policy debate about biotechnology has been couched in terms of potential risks to the environment and/or human health. If, however, people are more swayed by moral considerations, public concern

[5] Gaskell et al., "Europe Ambivalent on Biotechnology."
[6] Ibid.
[7] Ibid.

is unlikely to be alleviated by technically based reassurances and/or regulatory initiatives that deal exclusively with the avoidance of harm.[8]

Regrettably, the study does not enumerate or address the specific moral concerns that rendered the creation of transgenic animals for research morally unacceptable. Below in our discussion, I attempt to provide a plausible rational reconstruction of justifiable social moral concern about the production of transgenic animals for biomedical research and how this could be addressed. I also attempt to provide some reasons that xenotransplantation would be perceived as morally problematic.

The key point that we glean from Gaskell, however, is the degree to which the scientific community's neglect of ethics had an impact on the social acceptability of biotechnology. Despite these unmistakable results, I see no evidence that scientists are embracing discussions of the ethics of biotechnology, and both regulators and scientists still seem to believe that risk is the key item at issue.

My involvement with ethical issues in biotechnology began in 1987, when I was approached by the organizers of the First International Conference on Genetic Engineering of Animals to deliver the banquet speech.[9] In my conference paper, my subsequent book on the ethical issues occasioned by genetic engineering of animals, *The Frankenstein Syndrome*,[10] and in many papers and talks, I have used the Frankenstein story as a vehicle for explaining the relevant issues to the public and to scientists, and the tactic worked very well. I deploy the same strategy here, and show how it works well as a tool for dissecting out the relevant (and spurious) issue from any biotechnological modality.

Specifically, I was to address the topic of social and moral issues raised by the advent of this new and powerful technology. Flattered, stimulated, challenged, and totally ignorant, I accepted, confident of my ability to rise to the occasion by standing on the shoulders of my predecessors. Unfortunately, a brief visit to the university library shattered my preconceptions – I had no predecessors! My talk, in its published version, would be the first paper ever done on this major topic.

[8] Ibid.
[9] Rollin, "The Frankenstein Thing."
[10] Rollin, *The Frankenstein Syndrome.*

Suddenly, I saw my task under a new and harsher light. The buck stopped – and started – with me. It was truly an academic's nightmare.

Seeking a purchase on the topic, I solicited dialogue from colleagues in my philosophy department. "Genetic engineering of animals," mused one such partner in discussion, "You're talking about the Frankenstein thing." His remark was largely ignored by me at first, as it seemed to me flippant and shallow. It was only later that I realized he had opened a portal into the issue by forthrightly expressing what in fact rises to most people's minds when genetic engineering is mentioned. A week after our conversation, I was perusing new acquisitions in our university library when I encountered an extraordinary, newly published, five-hundred-page volume entitled *The Frankenstein Catalogue: Being a Comprehensive History of Novels, Translations, Adaptations, Stories, Critical Works, Popular Articles, Series, Fumetti, Verse, Stage Plays, Films, Cartoons, Puppetry, Radio and Television Program, Comics, Satire and Humor Spoken and Musical Recordings, Tapes and Sheet Music Featuring Frankenstein's Monster and/or Descended from Mary Shelley's Novel,* appropriately authored by a man named Glut.[11] The book is precisely a comprehensive list and brief description of the works mentioned in the title. After recovering from my initial amazement that anyone would publish such a book, I was astonished anew by its content. It in fact lists 2,666 such works, including 145 editions of Mary Shelley's novel, the vast majority of which date from the mid-twentieth century. Putting these data together with my friend's remark, I experienced a flash of insight: Was it possible that the Frankenstein story was, in some sense, an archetypal myth, metaphor, or category that expresses deep concerns that trouble the modern mind? Could "the Frankenstein thing" provide a Rosetta stone for deciphering ethical and social concerns relevant to genetic engineering of life forms?

In the ensuing months, my hypothesis received support. While visiting Australia, I met with an animal researcher whose field was teratology: the study of birth defects, literally, the study of monsters. He had been extremely surprised to find that his work with animals had evoked significant public suspicion, hostility, and protest. "I can't understand it," he told me. "There was absolutely no pain or suffering endured by any of the animals. All I can think of is that it must have been the

[11] Glut, *The Frankenstein Catalogue.*

Frankenstein thing." And in its 1985 cover story on the fortieth anniversary of the Hiroshima bombing, *Time* magazine invoked the Frankenstein theme as a major voice in post–World War II popular culture, indicating that this theme was society's way of expressing its fear and horror of a science and technology that had unleashed the atomic bomb. And when ABC-TV made a documentary on genetic engineering, the program was entitled *The Real Frankenstein.*

In my ensuing discussion, I use the three aspects of the Frankenstein myth I developed in my book on genetic engineering to illustrate three sorts of ethical issues pertaining to biotechnology: its alleged intrinsic wrongness, its dangers, and its causing harm to sentient creatures.

The first aspect of the myth is best uttered in a comic-opera middle-European accent, ideally by Maria Ouspenskaya: "There are certain things humans were not mean to know or do." Concerns of this sort are an example of bad ethics driving good ethics out of circulation, since they cannot be parsed into rationally based genuine issues. This notion, which is indeed older than the Frankenstein story, in fact can be found in the Bible, when God proscribes (albeit not in that accent) the fruit of the Tree of Knowledge. Some theme such as this is often orchestrated in tabloid accounts of genetic engineering – the key notion is that there is something intrinsically wrong with genetic engineering: blurring species, "messing with nature," violating the sanctity of life, "playing God," and so forth. The major point is that the wrongness of the action is not alleged to be a function of pernicious results or negative utility or danger – it is just wrong. This kind of response seems to increase in direct proportion to the dramatic nature of the genetic intervention. If one is putting a human gene into an animal, or vice versa, or genetically creating "monsters," that is, creatures whose phenotype is markedly different from the parent stock, one is much likelier to occasion this response than if one is causing subtle genetic changes, for example, introducing the poll gene (for hornlessness) into cattle.

Nonetheless, the pervasiveness of this position cannot be overestimated. An Office of Technology Assessment survey showed that 46 percent of the public believes that "we have no business meddling with nature."[12] Similarly, a National Science Foundation survey showed that

[12] Office of Technology Assessment, *New Developments in Biotechnology*.

although members of the public generally oppose restrictions on scientific research, "a notable exception was the opposition to scientists creating new life forms. . . . Almost two-thirds of the public believe that studies in this area should not be pursued."[13] Though survey data are always suspect, these data certainly buttress our claim about the social importance of the first aspect of "the Frankenstein thing."

It is not difficult to find examples of such positions, but it is quite difficult to find arguments to buttress these examples. For the proponents of such a stance, the wrongness appears to be self-evident. But things that are allegedly self-evident in their wrongness are highly suspect – in my lifetime alone, I have heard vehement proclamations trumpeting the inherent wrongness of "racial mixing," of allowing women into male occupations, and of prohibiting corporal punishment of children. So, in examining these positions, one is obliged to try to reconstruct the positions rationally to see whether there is any possible coherent argument in their defense, even if such an argument has not been clearly articulated by opponents of genetic engineering. This is the strategy I adopt in discussing the first version of the Frankenstein thing, which claims that genetic engineering is inherently wrong.

Probably the most patent source of such a position is theological. When the first genetic manipulations of animals were effected in the early 1980s, for example, when the gene for human growth hormone was inserted into swine, a significant outcry arose from theologians and religious leaders.

The National Council of Churches has declared that genetic engineering of animals does not display proper respect for "the gift of life," a theme that pervades theological pronouncements on the issue.[14] In the same vein, twenty-four religious leaders issued a pronouncement against animal patenting, employing the following language: "The gift of life from God, in all its forms and species, should not be regarded solely as if it were a chemical product subject to genetic alteration and patentable for economic benefit."[15] Similarly, the same

[13] National Science Foundation, *Science Indicators*, 1988.

[14] M. Crawford, "Religious Groups Join Animal Patent Battle," *Science* 237 (1987): 480–81.

[15] Statement of 24 Religious Leaders against Animal Patenting.

statement asserts that "the combining of human genetic traits with animals... raises unique moral, ethical, and theological questions, such as the sanctity of human worth." In none of these pronouncements is any argument, exposition, or explanation provided, so presumably the truth of the position is seen as self-evident.

When one adopts a theological perspective, one can certainly understand the qualms that religious leaders might have about genetic engineering. The Judaeo-Christian tradition has been staunch in its belief that God created living things "each according to its own kind," with the clear implication that species are fixed, immutable, and clearly separated from one another. Nineteenth-century and contemporary opposition by religious factions to Darwin and Darwin's notion of the origin and flux of species illustrates the significance placed on fixed kinds by religious groups. For humans to meddle with species, possibly to create new species, to blur the lines between species, and, indeed, as Darwin did, to argue that humans and animals are continuous is to erode the special place of humans and to trade comfortable predictability and order for uncertainty.

More cynically, as one Catholic priest told me in the 1950s, "If humans start creating and radically changing life forms, one ultimate mystery which draws people to religion will disappear. Our hold on people will diminish, they won't need us as badly." In other words, humans will be as gods. The idea expressed by this priest, that God fills in holes where human knowledge is absent and must retreat in the fact of the growth of human knowledge, was wonderfully described by theologian Dietrich Bonhoeffer as the notion of "God of the gaps" and represents a primitive form of theology.[16]

Clearly, then, both traditional ideology and rational self-interest militate in favor of conservative church (or synagogue) opposition to genetic engineering. (It should be noted that not all theologians oppose genetic engineering. Some would affirm that it is part of "co-creation" between God and humans.) What is critical, however, is that such concern, even if well founded, does not represent a social moral issue. Advances in knowledge and technology that fly in the face of religious tenets may appear morally problematic to adherents to those tenets – many religious people were offended by Newton's

[16] Bonhoeffer, *Letters and Papers from Prison.*

physics or Darwin's biology – but that in itself does not create a moral problem for our secular society in general or for its social ethic.

The point is that merely theological concerns do not serve as a basis for asserting in the social ethic that genetic engineering is intrinsically, morally wrong. On the other hand, the fact that a concern is theologically based does not mean that concern has no moral import. To dismiss a concern without examination simply because it may be couched in theological language is to commit a version of the genetic fallacy – confusing the source of an idea with its validity. Indeed, if philosophers such as Dewey are correct, putatively religious concerns may well be metaphorical ways of expressing social moral concerns for which no other ready language exists.

Jeremy Rifkin, one of the earliest and most vocal secular critics of genetic engineering, has a different strategy for attacking genetic engineering as intrinsically wrong, one that appears to be couched in religious terms.

> Genetic engineers increasingly view life from the vantage point of the chemical composition at the genetic level. From this reductionist perspective, life is merely the aggregate representation of the chemicals that give rise to it and therefore they see no ethical problem whatsoever in transferring one, five, or a hundred genes from one species into the hereditary blueprint of another species. For they truly believe that they are only transferring chemicals coded in the genes and not anything unique to a specific animal. *By this kind of reasoning all of life becomes desacralized* [emphasis mine]. All of life becomes reduced to a chemical level and becomes available for manipulation.[17]

This notion of *desacralized* is pivotal here. Prima facie, this seems to be a theological notion; yet Rifkin operates from a secular perspective. So one can only guess at what he has in mind outside a religious context. Presumably, he means that one commits some sort of metaphysical transgression when one adopts the stance he deplores in the paragraph quoted above, some sort of secular sin, which has moral import. What this all means is difficult to determine; presumably, he is restating the very point at issue, namely, that it is intrinsically wrong to genetically engineer animals. The key concept of desacralization is left undefined or must be thought to be intuitively obvious.

[17] Rifkin, *Declaration of a Heretic*, p. 53.

One can deduce from Rifkin's argument that one transgresses against nature when one views life as a bunch of chemicals. Presumably, then, Rifkin is inveighing against the venerable position known as reductionism, which asserts in its epistemological version that all phenomena can be explained by appeal to the laws of physico-chemistry or, in its metaphysical version, that all natural objects are nothing but bundles of chemicals (or molecules or atoms). Reductionism in one form or another is as old as codified thought: Democritus, for example, the ancient atomist, asserted that only atoms moving about in space were "real"; all else, such as colors and tastes and smells and distinctions between living and nonliving, were "by convention" and thence ultimately unreal.

There is no question that a fundamental component of the twentieth-century scientific ideology I discussed earlier is a metaphysical and certainly epistemological commitment to reductionism. Positivism decreed that one did not have adequate explanations of phenomena until one had subsumed these phenomena under the laws of physico-chemistry, which provide us with precision and predictability. Thus, most contemporary scientists view molecular biology as more of a real science than organismic or ecological science, because it is expressible in physico-chemical terms, while the latter is not. And modern medicine has also grown increasingly reductionistic and concerned with the universal and repeatable substratum of disease (medicine as a science), rather than with its unique manifestations in individuals (medicine as an art).

Although reductionism has certainly achieved prominence in contemporary science, it is by no means unquestioned. Classic criticisms of a reductionistic approach were generated by Aristotle and have been reaffirmed by its contemporary critics. Aristotle pointed out that there were in fact four questions or types of explanation one could generate about a given phenomenon: What is it made of, what makes it happen, what is it, and what is its purpose. (He called these the material, efficient, formal, and final causes of a phenomenon, respectively.)[18] As a great defender of common sense, Aristotle pointed out that a good explanation contains all of these components and that reductionism arbitrarily limits itself to the material and efficient causes. One can

[18] Aristotle, *Metaphysics*, Bk I, Chapter 3.

certainly explain the flight of a bird, for example, in terms of atoms and forces and aerodynamics, but one then loses sight of the fact that what is flying is a bird (its "birdness," in his terms) and also of the fact that it is an animal flying for a purpose (e.g., to migrate or to breed). This debate between mechanistic reductionism, on the one hand, which emphasizes the uniformity of all phenomena and attempts quantitative explanations of all things by appeal to the same set of laws, and teleological materialism, on the other, which emphasizes qualitative differences, each explained by laws appropriate to their own domains, continues to rage. In truth, this debate will probably rage forever, for even if we can explain all human behavior, for example, in physicochemical terms, mechanically, without appeal to intentions or other such concepts, it is an open question whether we *should*, or whether doing so is adequate.

Thus, for example, I have argued elsewhere, as have many others, that an exclusively reductionistic approach to medicine and disease is highly pernicious. In the first place, it ignores the valuational basis underlying concepts of health and sickness, for value notions and "oughts" are not expressible in the language of physics and chemistry. In other words, as discussed earlier, when we call a set of physical conditions a disease that ought to be treated or a state of an organism healthy, that is, what ought to be aimed at, we are making implicit reference to value judgments. It is, after all, a value judgment to say, as contemporary medicine does, that obesity is a disease, rather than a condition conducive to disease; or that alcoholism is a disease, rather than weakness or badness; or that child abusers are sick, rather than evil; or that health is a "complete state of mental, physical, and social well-being," as the World Health Organization definition asserts. Indeed, one legitimate concern about the ability to genetically engineer traits into humans is that it can lead to questionable values informing our concept of health – if everyone can be made 6 feet 2 inches tall, mesomorphic, and blond, and we value this configuration of traits, anyone not exhibiting these traits may be viewed as defective or unhealthy, with a resultant tendency toward loss of diversity in the human population.

In the same vein, reductionistic medicine can be criticized as ignoring the fundamental fact that diseases manifest themselves differently in different individuals, a point dramatically illustrated by Oliver

Sacks regarding Parkinson's disease in his *Awakenings*.[19] Similarly, pain experience, even given the same lesion, can very enormously in many modalities across different humans, across different animals, and across different cultures. This loss of individual perspective – illustrated by physicians' tendencies to talk of "the kidney in Room 407" – can and surely does generate a failure to take account of individual differences, to the detriment of patients.

We may thus surely grant that reductionism may be metaphysically wrong (in attempting to ignore qualitative differences), epistemologically wrong (in allowing for too few types of explanations), and even morally wrong, insofar as it leads to pernicious ignoring of real individual differences. But does this allow us to say that genetic engineering of animals is inherently wrong?

I can see no logical basis for such an inference. Even if many genetic engineers are in fact reductionists, it does not follow that they must be. For one can hold a variety of perspectives on the nature of things and still engage in genetic engineering. An Aristotelian could wish to create qualitatively new species, for example – there is in fact a strain in Aristotle, less known than his view of fixed, immutable species, that suggests that creation contains or could contain an infinite spectrum of species. One can argue that organisms are more than just bundles of matter – for example, that new properties emerge when we recombine matter – yet still argue both that we can create new organisms simply by manipulating matter and that there is nothing wrong with doing so. By the same token, one could without absurdity be a reductionist and argue that current configurations of matter in motion are "sacred" in Rifkin's sense or even in a theological sense, and thus should not be tampered with. Such an argument might proceed as follows: God made everything out of the same fundamental stuff according to basic laws but did so in the best possible way, and thus we should preserve, not tamper. Newton might well have so argued.

Recall that we are looking in this discussion for an argument that fleshes out the view that genetic engineering is intrinsically wrong, regardless of the consequences that issue from it. We have thus far looked at Rifkin's implied suggestion that it is wrong because it is tied to reductionism. Even if reductionism is both incorrect and conducive

[19] Sacks, *Awakenings*.

to morally wrong actions, it does not follow that genetic engineering is wrong. To prove this, one would need to show that all reductionism is inherently morally wrong, rather than capable of leading or even likely to lead to bad consequences, and that genetic engineering is inherently connected to reductionism, which as I just indicated is surely not the case. Below I discuss the claim that genetic engineering is wrong because it is likely to lead to bad consequences, but that is quite different from the current question: Is genetic engineering of animals just morally wrong, regardless of consequences?

Let us return to our first attempt to extract a nontheological account of the intrinsic wrongness of genetic engineering of animals from Rifkin's claim quoted above, which I have argued is untenable. The quotation is rhetoric, not argument, and is essentially question begging. Indeed, it is not even coherent in Rifkin's own terms. In the paragraph quoted above, he decries the fact that for genetic engineers, "the important unit of life is no longer the organism, but rather the gene."[20] Yet one sentence earlier, he condemns them for asserting that "there is nothing particularly sacred about the concept of species."[21] A few paragraphs later, he takes these engineers to task as follows:

> What, then, is unique about the human gene pool? Nothing, if you view each species as merely the sum total of the chemicals coded in the individual genes that make it up. It is this radical new concept of life that legitimizes the idea of crossing all species' barriers, and undermines the inviolability of discrete, recognizable species in nature.[22]

The astute reader will have honed in on some flummery in the above or, at the very least, on some fundamental confusion. Is Rifkin bemoaning the fact that genetic engineers do not see the individual organism as the fundamental unit of life, as he sometimes asserts, or the fact that genetic engineers do not respect the species as the fundamental unit of nature? Obviously, both cannot be fundamental.

Indeed, in the space of one page, Rifkin commits himself to two fundamental, ancient, and incompatible philosophical positions: nominalism and realism. Nominalism is the view that the fundamental furniture of the universe consists, as Aristotle said, of unique, particular,

[20] Rifkin, *Declaration of a Heretic*, p. 53.
[21] Ibid.
[22] Ibid.

discrete, concrete, and spatially and temporally located individuals, "this here existent thing," and realism is the view that what is ultimately real are classes, abstract, Platonic entities, which are not located in space and time, species whose reality transcends the spatiotemporal reality of the particular individuals that instantiate the "essence" in question. In the realist view, individual chairs come and go, but the form or essence of "chairness" endures. There are many respectable nominalists and many respectable realists, but one cannot be both, any more than one can be both a bachelor and married. (As Rifkin tries valiantly but incoherently to be both in the area of metaphysics, so too do many males vis-à-vis their spousal state.)

In fact, the debate over the ultimate reality of species or individuals in biology, a special case of the nominalism/realism debate, is still extant in biology. Some scientists argue that only individuals are real; species are at best a convenient artifice for grouping individuals. Others equally vehemently argue that species are the fundamental units of biological reality – species endure and are knowable while individuals come and go. (Interestingly enough, though Aristotle himself favors the reality of individuals, he concludes that individuals are unknowable, for we can separate out essential features from incidental or accidental ones only when we have a large group to sift through. It is essential to the nature of humans that humans are rational, but how do we know whether it is essential or not to the nature of Groucho Marx that he smoke a cigar?) In fact, little in biology hangs on the debate – realist biologists must still examine individuals, and nominalist biologist still acknowledge species. But these different metaphysical positions do generate very different stances on a variety of bioethical issues, including the moral status of animals, environmental ethics, and genetic engineering, so the issue must be addressed.

As we shall see, one encounters the appeal to the inviolability of species in a variety of quarters. The illegitimacy of crossing species barriers, the inviolability of species, comes up again and again in Rifkin's writing and in the pronouncements of those others who see genetic engineering of animals as intrinsically wrong. So the question arises: What sense can be made of the notion of species inviolability? Our paradigm case for the wrongness of "violating" something in a morally relevant way comes from our consensus ethic for the treatment of human beings. We "violate" people by causing them physical or psychological harm, frightening them, humiliating them, stealing or destroying their possessions, thwarting their freedom, not allowing them to express their natures, violently intruding into their physical or emotional lives, and so forth. And this is considered immoral or wrong because what we do to humans matters to *them*; thwarting their interests and desires and needs makes them suffer. We have seen how society is plausibly extending this notion to animals, where we have good reason to believe that what we do to them matters to them. On the other hand, we cannot violate – or hurt – a rock or a tree except metaphorically, since what we do to nonsentient entities doesn't matter to them. To be sure, we can "violate" a person's house or car, as when we force our way into them and do damage, but we are harming the person who owns the house or car, or those who could use them, not the objects in a morally relevant sense. We cannot wrong the car.

Similarly, can I harm or violate a species? Humans can decimate a species, like the buffalo, or destroy it completely, as in the case of the passenger pigeon, but have we done harm to the species? In my view, we have harmed in a morally relevant way animals who comprise the species; or perhaps the humans who depend on those animals or plants or who admire them aesthetically; or the humans who care about biodiversity; or else the animals that depend on the members of the vanished species. But we have not harmed the species, because a species is not sentient; only the members of some species are. A species, if it has any existence above and beyond the members thereof, exists as an abstract Platonic entity like a set or a number. We do not harm the number one by destroying all the things in the universe it can denote; for a Platonist, it continues to endure in splendid isolation. By the same token, the class (or species) of buffalo does not care whether it has few or many members, though the buffalo presumably do.

To be sure, all right-thinking people are or ought to be greatly disturbed when a species, whether it consists of sentient members or not, is threatened with the extinction of its members. Why? Because dodoes, or Siberian tigers, or an endangered moss will never pass this way again; because they may well be inextricably linked with the survival of other sets of individuals that we or others care about or, indeed, with our own survival; because they are beautiful, and we want our children and ourselves to be able to experience them. But we can also care about the defacing of the Mona Lisa or the erosion of the Sistine Chapel without granting them moral standing in themselves, but strictly on the basis of their relations with sentient beings. As far as we are concerned, it may be far worse to kill the last ten Siberian tigers than ten other Siberian tigers when there are many of them. (This is, of course, the basis of the conservation ethic.) But as far as the tigers are concerned, they don't know or care whether they are the last, and thus it is equally wrong from a tiger's perspective (if it is wrong) to kill any ten or the last ten. And as I said earlier, none of it matters to the species at all, for things do not matter to species, and, if one is a realist, species continue to exist whether or not their members do.

The mistaken tendency to identify species as sacred things has many sources. First and foremost, as we have already seen, is the biblical notion that God created kinds. And if we annihilate – or otherwise meddle with – kinds, we are meddling with God, to whom, as the Bible indicates, our actions matter a great deal. Second, the dominant version of Aristotelianism that has pervaded Western culture also stresses the fixed and immutable nature of things. Aristotle's reasons for believing in fixed and immutable species seem to come from two fundamental sources. In the first place, as a common-sense philosopher, Aristotle points out that if species changed, we would surely see cases of such transmutation, yet this has never been noted by human beings. (Aristotle and most of his contemporaries did not view fossils as evidence of vanished beings – they rather saw them as kinds in themselves. Thus fossil fish were stone fish, not traces of bygone fish.) Second, again operating from a common-sense basis, Aristotle points out that if natural kinds changed, knowledge would be impossible, since knowledge entails stability, and nothing is plainer than the fact that humans are by nature "knowing creatures." Thus, both common sense and Aristotle's articulation thereof loom large in Western culture's tendency to see natural kinds as fixed.

Yet another factor militating in favor of viewing species as fundamental units of reality has come from the scientific community: Significant portions of the biological scientific community have argued that species are in some sense "more real" than other units of classification. In other words, in this view, species concepts are more than arbitrary, Dewey-decimal-system sorts of classificatory schemes, but rather somehow accurately subtend the way things are "in the world." Though biological taxonomists have differed as to the way in which species are characterized – some, for example, have located the nature of species in genetic similarity, others in genealogical or evolutionary history – both approaches would agree that species locutions are closer to reality than the other classificatory concepts commonly employed, which are more inclusive than species. These are, as any biology book recounts, in decreasing order of generality, kingdom, phylum, class, order, family, genus, species. Thus, the domestic dog belongs to the animal kingdom, phylum of chordates, class of mammals, order of carnivores, family of canids, genus of *Canis*, and species *familiaris*.

Michael Ruse, a philosopher of biology, has succinctly summarized the view of biologists regarding the reality of species as follows:

> There seems to be common agreement amongst biologists... that there is something rather special about the biological species.... Somehow, groups which are biological species are felt in some sense to be "real" in a way that other groups are not felt to be.... Almost without exception, evolutionary taxonomists are adamant in their contention that it is biological species alone which are real.[23]

Interestingly enough, the root of this idea is probably in part common sense. Ernst Mayr, the classic defender of the reality of species, argues that the reality of species is apparent in the universal way in which humans delineate groups of animals and plants as separate from one another:

> The primitive Papuan of the mountains of New Guinea recognizes as species exactly the same natural units that are called species by the museum ornithologist. The arrangement of organic life into well-defined units is universal, and it is this striking discontinuity between local populations which impressed the naturalists Ray and Linnaeus and led to the development of the species concept. There can be no argument as to the objective reality of the gaps between local species in sexually reproducing organisms.[24]

This sort of approach is indeed reflected in the standard biology textbook account of species; for example, one recent text asserts that a species is "a genetically distinctive group of natural populations... that

[23] Ruse, *Philosophy of Biology*, p. 127.
[24] Mayr, "The Species Concept," p. 371.

share a common gene pool and that are reproductively isolated from all other such groups."[25] Another text asserts that a species is a group whose "members can breed successfully with each other but not with organisms of another species."[26] (We can charitably forgive the patent circularity in the above definition, as it does not interfere with our point.) Still other definitions add that members of a species must be able to both successfully interbreed and produce fertile offspring – this qualification makes clear that animals such as horses and donkeys are not in the same species, as they can and do successfully reproduce, but they do not produce fertile offspring (mules being infertile). On the other hand, the definition is still not adequate, as there are in fact groups we call separate species that breed and produce fertile offspring – dogs and wolves provide a familiar example, as do dogs and coyotes; lions and tigers can also be bred to produce fertile offspring. By the same token, Chihuahuas and Great Danes are considered members of the same species, yet they are clearly unable to breed without human help and thus do not meet the definition of species.

It is widely held in biology that there is, in fact, no adequate definition of species that is not subject to refutation by counterexample. The notion of species thus appears to be a fuzzy concept, one that is not precise or definable but that captures our intuitive tendencies about grouping what we find in the world. And given the evolutionary paradigm, species are going to have very fuzzy boundaries, as new species emerge from old species by new selection pressures that favor changes that have emerged by chance.

Thus modern biology does not accept the notion of species as fixed, immutable kinds, a notion biblically enunciated and articulated in Aristotle. In post-Darwinian biology, nature is forever experimenting with modifications of extant species, most of which modifications will prove to be deleterious, but a small number of which will contribute to the incremental (or, according to some theorists, dramatic) changes that eventually produce new species. One can therefore argue that, if species are not fixed, as far as nature is concerned, and humans are part of nature, there is no reason to say that humans cannot produce new species – we have certainly done so unwittingly in countless

[25] Keeton and Gould, *Biological Science,* p. 883.
[26] Mader, *Biology,* p. 745.

cases by radically altering environments. Indeed, we have caused the extinction of countless species in the course of human civilization and, as environmentally concerned scientists point out, are doing so now at an unprecedented rate. In fact, it may well be that concerns about annihilation of species at human hands spill over into concerns about changing species – it is a small step, psychologically, though logically an indefensible one, to go from the claim that species ought not to become extinct due to human actions to the claim that species ought not to change at human hands.

We have certainly drastically altered species in an intentional way, ever since we have been able to do so. And we have in fact created new plant species in abundance through hybridization – the tangelo and some types of orchid are mundane examples of such genetic manipulation. Indeed, it is estimated that 70 percent of grasses and 40 percent of flowering plants represent new species created by humans through hybridization, cultivation, preferential propagations, and other means of artificial selection.[27] Few people feel that such breaching of species barriers is intrinsically morally problematic. On the other hand, though we have produced mules, beefalos, tiglons, ligers, coydogs, and other sterile and nonsterile hybrids in the animal kingdom, we have never been able to produce a new species, in the sense of animals so far removed from the parent stock that they are reproductively isolated from it but still fertile. Nor have we yet done so via genetic engineering; nevertheless, this is clearly what drives concern of the sort enunciated by Rifkin, for it is surely in principle possible, though not practically possible given existing technology.

So it appears that, for the immediate future at least, genetic modification of animals will not breach the species barrier but will accelerate the sorts of modifications that we have hitherto effected using artificial selection. Breeding for phenotypically evident traits such as greater size, quicker growth, and disease resistance does not involve creating new species and does not differ in kind from what we have always done. It does, however, differ in the rapidity with which changes can be introduced into a genome; as one of my colleagues put it, current techniques of genetic engineering allow us to produce changes "in the fast lane." This in turn means that we do not go through a long

[27] D. Pettus, personal communication.

period of trial and error over many generations during which time we can detect untoward consequences of our engineering that can cause harm to the animal or to humans or to other animals. But this in turn does not provide a reason for believing that genetic engineering is intrinsically wrong, while traditional manipulation of genomes through breeding is not. Rather, it tells us that genetic engineering may occasion undesirable consequences. Perhaps the risk is so great as to militate against manipulating animals in the fast lane – I address the dangers of genetic engineering below – but then the wrongness of genetic engineering is not intrinsic, but consequential. And we are here attempting to unpack the possible arguments for its allegedly intrinsic wrongness.

Another possible source of the belief that genetic engineering is inherently wrong is environmental philosophy, a creature of the 1970s that arose in response to new and alarming information about the degree to which human activities were wreaking havoc with the environment. From Rachel Carson's accounts of the devastating effects of pesticides to acid rain to visible air pollution to the amazing fact that one could set on fire the industrially polluted Cuyahoga River in Ohio by dropping a match into it, we were galvanized into significant concern about the degree to which we were harming nature and into thinking about our relationship with nature. Now there is no question that ethical thinking relevant to how we approach our environment has been sorely lacking in human history, at least in technological societies. So-called primitive societies have long realized that respect for the natural world was a necessary condition for survival, and such respect was mirrored in their religions, mythologies, and social organization. A complex example of such a culture may be found in Australian aboriginal society, highly stable in its fundamental tenets for millennia. Such harmonious stances were clearly born of necessity; the price of a cavalier stance toward nature was a precipitous diminishing of one's chances of survival.

With the rise of technology, however, there came an apparent ability to tame or harness nature. When this was coupled with the strong Christian tendency to deny any sacred status to nature (in part to distance itself from autochthonous religions) and to view nature as provided by God for human use, the stage was set for environmental despoliation. With the advent of the industrial revolution's "dark satanic mills" came fouling of air and water and soil on an ever-increasing scale. Species of plants and animals were driven to near extinction and extinction to provide for all manner of human desires and to create wealth from "natural resources." As technology grew exponentially and urbanization increased, people grew less sophisticated in their understanding of human dependence on nature. With this came an attitude that humans were improving the natural world, adding value to a limitless supply of raw material. And with this came also a new hubris, that we could improve on nature endlessly at little or no cost, a stance that well characterizes our ever-increasing dependence on agri-chemicals and the belief that nature was not only an endless source of bounty but an endless and forgiving receptacle for the waste products of high technology. Indeed, I distinctly recall my mind-set and that of my peers in the early 1960s: an unlimited confidence in science and technology positively to dominate and control and overcome nature at little or no cost.

In my own case, which I suspect is rather typical, I was "shocked out of my dogmatic slumber," in Kant's felicitous phrase, by a series of related incidents. The first occurred when, in 1962, I was caught in the rain one beautiful spring day in New York City and found that wherever my bare skin had been exposed to the shower, the skin was burning, itching, and inflamed. On rushing to my physician, I found that I had experienced an extreme reaction to acid rain, resulting from air pollution. Three years later, the same polluted air was driving me to the emergency room of St. Luke's Hospital on a regular basis, often three or more times in a week, the victim of chronic asthma. By 1967, I was quite conscious of issues of air quality and was appalled to find, when touring remote portions of northern Canada, that the people living there, under subsistence conditions, were afflicted with extreme air pollution from U.S. Midwestern factories, while at the same time reaping none of the benefits emerging from that industry.

Thus when the Phi Beta Kappa chapter at the City College of New York surveyed its members in the mid-1960s on what they considered the most significant problems facing American society, I was the only one to put environmental concerns at the top of the list. By the early 1970s, after the first Earth Day, this had changed substantially. Today the vast majority of Americans consider themselves environmentalists, and even kindergarten children worry about Mother Earth, recycling, and the devastation of the rain forests; terms such as *biodiversity* are household words.

This is, of course, all to the good. For reasons that are not clear, however, a group of environmentally concerned philosophers have felt compelled to generate a radically "new ethic" for the environment and have argued that natural objects (concrete and abstract) – ecosystems, rivers, species, and nature itself – possess intrinsic or inherent moral value and are direct objects of moral concern to which we have moral obligations. Indeed, for many of these theorists, these entities have higher value and more inherent worth than "mere individuals," human or animal. By arguing this case, such theorists hope to introduce a solid grounding for environmental concern. This, in turn, tends to foster a "nature is perfect as it is" attitude, which naturally fuels aversion to changing species or making other modifications by genetic engineering and, indeed, suggests that this is morally wrong. Both the environmentalism that has become pervasive in society and environmental philosophy probably contribute to the mind-set that views the genetic engineering of animals (or, for that matter, of anything – plants and microbes included) as intrinsically wrong. Social environmental thought is concerned that species not disappear at human hands, holding that it is just wrong (intrinsically wrong) for us to destroy species. (I would argue, as above, that it is *instrumentally* wrong.) As we remarked earlier, it is psychologically a small step (but logically a vast one) to move from the claim that species should not vanish at human hands to the more dubious claim that species should not change at human hands.

As far as environmental philosophy is concerned, the specter of God, or reified and deified evolution, lurks behind some of the theorists – Holmes Rolston, for example. They say that if nature is the product of God's (or something's) infinite wisdom, and if it is intrinsically valuable, and if it is a great systemic whole, it is easy to deduce that we, with our pitiful intellects, should not tinker with species, for we cannot alter a part without altering the whole.

Not surprisingly, Rolston quite consciously reifies species. He asserts that "species *is* a bigger event than individual interests or sentience."[28] He further refers to destruction of species as "superkilling," which is

[28] Rolston, "Duties to Endangered Species," p. 213.

morally worse than the killing of individuals – even the same number of individuals if they didn't exhaust the species.[29] Furthermore, he argues that we are morally bound not merely to preserve species but to preserve them in the natural, biological, evolutionary-ecological context in which they developed.

> A species is what it is inseparably from its environment.... It is not preservation of *species* but of *species in the system* that we desire.... The full integrity of the species must be integrated into the ecosystem. Ex situ preservation, while it may save resources and souvenirs, does not preserve the generative process intact. Again, the appropriate survival unit is the appropriate level of moral concern.[30]

Such a position, of course, strongly disposes one against genetic engineering – our wisdom cannot begin to compare to natural wisdom. So Rolston would presumably clearly resist genetic engineering of anything that could conceivably affect nature, though he does not address this question in his writings. Furthermore, even though he applauds the proliferation of species in nature as intrinsically valuable, this would presumably not apply to humanly created species, since they would be artificially imposed on nature, not be "projected" by nature, and thus would not fit "in the system." It is not clear, however, that Rolston would oppose genetic engineering of domestic animals, which he describes as "captured in culture,"[31] and thus sees as radically distinct from nature and not covered by our moral obligations to "natural species."

In conclusion, note that I am not, of course, despite my criticisms of environmental philosophy, suggesting that we should be cavalier about altering nature. I am merely suggesting that circumspection in this area – including circumspection about genetic engineering – should be not a result of the intrinsic wrongness of such alteration but of the possibility that untoward consequences that might result. But it is worth noting that humans have been altering nature since they crawled out of the primordial ooze; for better or worse, it is what we do, even as fish swim and birds fly.

[29] Ibid., pp. 212–13.
[30] Ibid., pp. 215–16.
[31] Rolston, *Environmental Ethics*, p. 59.

In my *Frankenstein Syndrome* I have argued at great length that philosophical environmental ethics is overkill, violating Occam's razor. For an adequate environmental ethic, one needs only the prudential maxim, "Don't foul your own nest." The argument that nature has intrinsic value is not only incomprehensible, but it is not helpful, for humans are the paradigm instances of "intrinsic value," and to ascribe it to inanimate objects just pits one form of alleged intrinsic value against another, with no way of weighing them. In my view only sentient beings can have intrinsic value anyway, because the only sense I can make of intrinsic value is that what happens to you matters to you, even if no one else cares.

We have devoted a great deal of attention to the claim that genetic engineering is "just wrong" or intrinsically wrong. This is because these claims have the strongest hold on the social mind and are very difficult to refute. We have found that such claims, when unpacked, fall into one of two categories. Either they involve religious appeals that cannot be translated into secular moral terms, appeals to portentous but vacuous notions, such as "desacralizing nature" or "breaching species barriers," and appeals to questionable metaphysical categories, or they turn out to be claims that genetic engineering is wrong because it will have bad consequences or cause great harm. The arguments for the former category can be put behind us, but the claim that genetic engineering of animals is wrong because it can or will inevitably generate bad consequences, though very different from the claim that genetic engineering is intrinsically wrong, is highly troubling. Indeed, all cogent accounts of the first version of "the Frankenstein thing" come down to arguments about danger to humans, animals, or nature that could eventuate from genetic engineering. Thus, from the first version of "the Frankenstein thing" we must shortly turn to the second, the idea that genetic engineering of animals, although not wrong by its nature, is wrong because it is likely to produce significant harm. This thesis occupies our concern in the next chapter.

There is another concern sometimes voiced by those who allege the intrinsic wrongness of genetic engineering of animals. Theological types, in particular, as we saw earlier, object to the mixing of human and animal traits. Presumably, this means the insertion of human genetic material into animals or the insertion of animal genetic material into humans. The former has occurred; as I discuss below, the human growth hormone gene has been inserted into animals in order to create animals who grow faster and leaner. To my knowledge, the latter has not been attempted.

What, precisely, is intrinsically wrong with such an admixture? We have certainly inserted animals' parts into the human body to treat disease – pig heart valves and pig skin, for example – and routinely use animal products for medicinal purposes. Suppose an animal were found that contained genes for preventing cancer – sharks, allegedly, are tumor-free. Suppose, further, that this gene produced no untoward effects in humans; indeed, its only effect was to confer immunity against neoplastic disease. Why would such gene transfer be wrong in and of itself? Similarly, if the case were reversed and the gene were transferred in the other direction, from humans to animals to instill disease resistance, it is difficult to see why this would be morally problematic. To be sure, when the human growth hormone gene was transferred to animals, it caused animal disease and suffering, and thus such transfer was wrong because of its effects. But one has still not shown that in and of itself such transfer is wrong.

Below I raise the interesting case of transferring the genes that cause human genetic disease to animals in order to create models for human genetic diseases, which models can easily be studied and upon which experimental manipulations can be done. Many people see this as a way eventually to cure human genetic disease. Others feel that it is wrong because of the vast amounts of pain and suffering the animals will undergo if the transfer is successful. But once again, the moral dimension of this debate seems to arise when one considers the consequences of genetic transfer, not the transfer in and of itself. Below I also discuss xenotransplantation – genetically altering animal organs for transplantation to humans.

7

Biotechnology and Ethics II

Rampaging Monsters and Suffering Animals

The statement "There are certain things humans were not meant to know (or do)" often conceals an enthymematic component "because such knowledge will lead to great harm." This theme is orchestrated in the tales of the Tower of Babel, the Sorcerer's apprentice, the Rabbis who studied the Kabbalah and went mad or turned apostate, and of course the endless variations on the Frankenstein story we alluded to earlier.

The reason that disaster results in these stories, of course, is not so much that a given area of knowledge or its application is inexorably dangerous, it is rather that we (or scientists) tend to rush headlong into a field or activity with incomplete knowledge, where our ignorance inexorably leads to disaster. And we have seen abundant examples of this in twentieth-century science: the escape of "killer" bees, the Chernobyl and Three Mile Island disasters; the various space shuttle tragedies. The thrill inherent in the pursuit of new knowledge or new power often eclipses scientists' concern about dangers – I have seen scientists engrossed in an experiment use their mouths to pipette drug-resistant tuberculosis!

There is in fact a tendency on the part of researchers to denigrate the need for any biosafety oversight; "I've always done it this way and no one has ever been hurt" is a familiar refrain. On one occasion, before it was mandatory for research institutions to have a biosafety officer, I was discussing these issues with the provost of a major research university. I was arguing that, mandated or not, having a biosafety officer charged with enforcing compliance with sound principles of biosafety was a fundamental moral obligation of any institution undertaking research. "What are you worried about?" I was told. "Nobody's been killed here yet."

A final example, directly germane to genetic engineering, will round out my point. A few years ago, I was invited to lecture on genetic engineering at a major research university that had just received $50 million from its state legislature for developing biotechnology. The funding was largely the result of the efforts of two state legislators, both farmers, who felt that strengthening biotechnology was required for the state to compete economically. The two legislators were wise; they had written into the funding the proviso that a certain percentage of the money had to be used for examining social and ethical issues engendered by biotechnology. After my talk, I attended a party for the speakers, where the two legislators were also present. One of them was visibly angry, and I immediately assumed that my talk had provoked that rage, as my talks so often do. He assured me that this was not the case and that his ire was in fact directed against the scientists in charge of the program. "Why?" I asked. "Didn't you hear his speech?" he replied. "His very first project involves working with a chemical company to test their new herbicide-resistant seeds. The theory is that one can soak the earth with their herbicide, yet the seeds will grow." "So what?" I asked naïvely. "So what! We happen to have a ground-water contamination problem here already as a result of herbicide overuse. This will certainly add to the problem!" I began to understand his concern. "Have you approached the scientist?" I asked. "How on earth can he justify such research?" "That's why I'm so angry," said the legislator. "He said there is no cause for concern, that they would fix the problem by genetically engineering a microbe to eat the contaminants." Suddenly full understanding dawned on me. "I see," I said. "You're worried that it will also eat the state capital, and if you express this, he'll say 'Don't worry, if that happens we'll build a microorganism to puke the state capital.'"

In sum, then, scientists tend to be cavalier about the dangers emerging from science and technology, as Dr. Frankenstein was, or tend to ignore them, and this lends credence to a second version of "the Frankenstein thing," for scientists are unlikely to anticipate potential dangers on their own.

Ordinary people are by and large wary of expert assurances that newly discovered areas of research or applications of research are free of risk. Many of us have witnessed numerous cases in which scientists' reassurances about the safety of new innovations and the minimal dangers to society of such innovations have turned to ashes. Experts told us that living near a nuclear plant is safer than taking a bath – this was dramatically belied by Three Mile Island and, more tragically, by Chernobyl. Experts told us that "there is no scientific evidence" that living near electrical transmission lines poses a danger to health, yet reports of the high incidence of leukemia in children who live near such lines continue to roll in.

We can all recall numerous examples of failure of expert reassurance or cases where experts have blundered or been blindered. Who among us can forget the space shuttle tragedy, when the shuttle was launched on a freezing cold day, and all the alleged fail-safe mechanisms failed? Who among us has not known a tragic misdiagnosis – women's shoulder pain dismissed as "menopause" and turning out to be lung cancer or heart disease; men's back pain dismissed as a charley horse and turning out to signal prostate cancer – despite our formidable armamentarium of diagnostic tools?

In the early 1980s, I personally experienced an incident of expert failure – failure to acknowledge the reality of a risk and the correlative dissemination of reassurance to a community that it was not at risk. When our university was contemplating the undertaking of AIDS research, members of the community were fearful of the possible risks to the general population that might emerge from such research. As it happens, there is a branch of the Centers for Disease Control (CDC) located at our university, and when a major AIDS expert from another CDC research center came to the community, he agreed to speak to the public about the risks stemming from researching the virus. Such risks, he assured a large audience, were minimal – virtually nonexistent. One needed "to work" he said, at catching the virus. Outside of engaging in venereal contact with an AIDS sufferer or receiving a transfusion of infected blood, one simply could not contract the disease, he averred. There was, he said, "no scientific evidence that you can acquire AIDS in any other way." There was "no risk" of acquiring the disease in a laboratory context. "Why," he declared, "I would bathe in the AIDS virus – there is no danger of catching it that way." Within six months of his skeptically received pep talk, the first case of a person contracting AIDS through a cut in the skin was reported. Obviously, the public reluctance to trust this expert was well founded.

In addition, author Michael Crichton has persuasively (and entertainingly) argued in his *Jurassic Park*[1] that risks associated with a new and complicated technology such as genetic engineering are inherently unpredictable, and thus all safeguards built into the system in advance are likely to fail.

Crichton's fictional vehicle for making his points is the prospect of recreating dinosaurs by genetic engineering for a dinosaur wildlife park. Although scientists build into both the animals and park various clever mechanism to prevent the animals from reproducing or leaving the confines of the preserve, things do not go as planned. The conceptual point underlying Crichton's story is drawn from the relatively new branch of mathematics known as chaos theory, which postulates that the sorts of intrinsically predictable systems beloved by Newtonians, determinists, and introductory philosophy professors – the billiard table whereupon, if all forces applied to the balls are known and the initial conditions are described, one can predict with certainty the movement of the balls – are few and far between and, in any case, are essentially irrelevant to complex new technologies such as genetic engineering. Crichton puts his point in the mouth of a mathematician named Malcolm.

"Jesus," Gennaro said. "All I want to know is why you think Hammond's island [the dinosaur preserve] can't work."

"I understand," Malcolm said.... "Simple systems can produce complex behavior. For example, pool balls. You hit a pool ball, and it starts to carom off the sides of the table. In theory, that's a fairly simple system, almost a Newtonian system. Since you can know the force imparted to the ball, and the mass of the ball, and you can calculate the angles at which it will strike the walls, you can predict the behavior of the ball far into the future, as it keeps bouncing from side to side. You could predict where it will end up three hours from now, in theory.

"But in fact," Malcolm said, "it turns out you can't predict more than a few seconds into the future. Because almost immediately very small effects – imperfections in the surface of the ball, tiny indentations in the wood of the table – start to make a difference. And it doesn't take long before they

[1] Crichton, *Jurassic Park.*

overpower your careful calculations. So it turns out that this simple system of a pool ball on a table has unpredictable behavior."

"And Hammond's project... is another apparently simple system – animals within a zoo environment – that will eventually show unpredictable behavior.... The details don't matter. Theory tells me that the island will quickly proceed to behave in unpredictable fashion.... There is a problem with that island. It is an accident waiting to happen."[2]

Chaos theory in fact represents a formalization of a number of points well known to common sense but that often escape the common sense of science. First, as any mechanic will readily attest, "things don't work like they're supposed to." There is a major gap between theory and practice. Blame it on gremlins, Murphy's Law, or just that "Shit happens," as a popular bumper sticker reminds us. Second, insofar as any system must be implemented by humans, a weak and unpredictable link is ipso facto present. Third, when scientists talk about Newtonian predictability, they are talking about closed systems where everything is tightly controlled, all relevant variables are known, and variations in initial conditions – imperfections in the pool table, in the above example – are abstracted from. In the real world, this does not occur. Not only are there imperfections in the table, but windows fly open and gusts of wind blow the balls askew, cockroaches crawl up on the table and block the movement of the ball, demented snipers shoot the ball, cats attack the ball, ceilings collapse, and so forth.

It is precisely this sort of insight that leads those with common sense to take what experts say with a grain of salt. "I don't care what the experts say" is often the response of the ordinary person living in an atmosphere that smells funny; "but breathing that stuff can't possibly do you any good."

Part of the conflict between science and common sense in this sort of arena has to do with differences in canons of evidence and justification – the scientist will never say that the funny smell is harmful until he or she has done exhaustive epidemiological work to confirm that there is a problem and has confirmed the operative mechanism that causes the problem. Common sense – most certainly a mechanism for survival – demands far less and is content with affirming on the basis of vague experience of the race that "what don't smell right ain't right."

[2] Ibid., pp. 76–77.

Scientists make careful, complex judgments, and they attempt to do this in a detached way; common sense's judgments have no pretense of disinterestedness. Here Hume is correct; one must work to be a philosopher or scientist, but one's humanity comes naturally. Thus, ordinary people remain fearful of buildings in which disease research is performed – especially on nightmarish pathogens such as the AIDS virus, anthrax, small pox, the tubercle bacillus (TB), and the rabies virus – despite scientific statements that adherence to federally mandated guidelines for containment assures that there is no danger.

Strictly speaking, avoiding disastrous outcomes of genetic engineering and other biotechnology itself represents not an ethical issue, but merely a prudential one. No genetic engineer – even one totally dominated by scientific ideology – wants to see disaster occur as a result of his or her work, not only out of a desire to protect the innocent, but even more so, as the Asilomar Conference imposing self-regulation on the burgeoning area of recombinant DNA showed, out of a fear that if something went wrong, a whole area of research might be closed down.

The ethical component in controlling risk rather comes in when one weighs possible risk against possible benefits. The best way to explain this is by way of a personal anecdote. In 1999, I was invited to participate in the World Health Organization policy-forming meeting on antibiotics in animal feeds. In the course of my presentation,[3] I argued that the U.S. government's own statistics show that abandoning antibiotics would raise food costs about $10 a year per household. I then argued that any parent would happily pay $10 a year to ensure against their children developing infections caused by antibiotic-resistant pathogens of the sort likely to emerge from unrestricted antibiotic use in agriculture. When I was finished, the FDA veterinarian in charge of this issue for the United States jumped up, white with rage. "I am offended!" she declared. "At what?" I queried. "At the presence of an ethics paper at a scientific conference! This issue is one of science, not ethics!" Patiently (or at least not as impatiently as I felt), I said to her: "The issue in question is weighing risk/benefit, is it not?" She agreed. "Imagine I give you enough research money to do as much research as you wish, so that you find out that using

[3] Rollin, "Ethics, Science, and Antimicrobial Resistance."

antibiotics in feeds kills (or makes sick) one person in 6 million, or 5 million, or 4 million or 300,000. Does this tell us which percentage of death or illness is acceptable? Of course not; that is a matter of ethics, not science." She seemed to get the point and was quiet for the rest of the conference. I heard later from some colleagues at Purdue that for months after, her supervisor went around to conferences on these issues assuring participants that the issue of antibiotics in feeds was *both* an issue of science *and* of ethics.

The relevance of this to our second category of ethical issue in genetic engineering – possible dangers – should be clear. Scientists and biotechnologists may feel that (for example) the possibility of one death in a thousand is a reasonable risk in exchange for the benefit of some new biotechnological modality. Ordinary people, with no vested interest in biotechnology, may demand no greater risk than one in a million. Here is where the ethical difference arises. For this reason, I have argued that the public will never accept biotechnology until it has a say in determining what counts as acceptable risks undertaken for possible benefits. I have further argued for such decisions being made democratically by local committees chartered in the area in which the research is to be done. Thus far, the biotechnology community and regulatory community have been very high-handed with regard to social ethics. An excellent example of this is FDA's (and biotech companies') unwillingness to label genetically modified foodstuffs, on the grounds that the products are identical and are merely derived from a different process.

This decision was reached with great arrogance – no public hearings were held, no attempt was made to solicit public input in any form. This in turn feeds the common-sense view that scientists and government have no regard for public concerns, have their own agenda, and are removed from ordinary common sense and its values. To announce blithely and without public discussion that tomatoes containing flounder genes are just like other tomatoes is to display either great ignorance of ordinary ways of thinking or great contempt for them, or both.

For those of us who follow biotechnology, this case inspired a sense of déjà vu. Similar insensitivity was displayed by the biotechnology industry when one of its first, and most publicized, attempts to market biotechnology products was the animal drug BST (bovine somatotropin), also known as BGH (bovine growth hormone). This product was basically designed to be injected into cows once a day during the experimental work on it, with an eye toward eventually marketing a product that would be injected every two weeks, in order to produce greater milk yield, thereby making diary farms more efficient.(BST works by reapportioning nutrients into milk secretion rather than fat deposition in the animal.)

The industry could not have picked a worse inaugural product if it had left the choice to Jeremy Rifkin. For BST raised public ire in virtually every conceivable segment of society. In the first place, issues of adulteration and food safety surfaced quite quickly. Despite assurances from the FDA and the biotechnology community of the safety and wholesomeness of the product, public confidence did not materialize. Other experts questioned the safety of the product, and successful consumer boycotts were initiated. In Britain, also, much publicity was garnered by those concerned about food safety.

In retrospect, the industry's willingness to press forward a product that would predictably raise questions about the safety of milk, a paradigmatic symbol of purity and wholesomeness, is astounding. It bespeaks either arrogance or stupidity. One need not be a Nostradamus to realize that parents would surely reject a product concerning which even a suspicion of danger of adulteration existed. But that was not the only folly manifest in this case. The industry should surely have realized that a public accustomed to cheap milk and to surplus milk – weaned, as it were, on picture of farmers dumping milk to protest low prices and of warehouses full of surplus dairy products – was not likely to be overly sympathetic to a new product that increases milk yield.

Finally, BST stirred the wrath of small dairy farmers and of the multitude of citizens for whom "the family farm" is, as it were, a sacred cow. Farmers argued that if BST came into common use, small dairymen would not be able to afford its costs, would not be able to compete with large corporate dairies, and would thus be driven out of business. Advocates affirmed that BST is neutral with regard to the size of the operation, that is, does not favor large producers.

As if all this were not enough, issues regarding the welfare of the cows to be treated with BST were also raised. It is well known that modern dairy cows, while producing significantly more milk than their predecessors, also have a significantly shorter productive life. Whereas traditional milk cows might continue to produce for an average of three to four years, today's high producers can be culled for metabolic diseases within two to three years, and thus lose their value and get shipped to slaughter. BST, it was argued by some, would only increase this tendency. In addition, BST might also increase "production diseases," that is, those diseases associated with the stress of high productivity. BST-treated cows do, in fact, show a higher incidence of mastitis than do untreated cows.

Again, the key point, for our purposes, is that the industry was apparently blindsided by these objections. This should not have been the case. Any reasonably intelligent citizen, if informed about BST, was likely to have raised some or all of these objections. Certainly, small dairy farmers would have been able to do so. And yet this didn't happen. So one can assume that the industry (i.e., those hoping to market BST) felt no need to worry about the social reception of the product – a degree of short-sightedness that was again reflected in the FDA decision just mentioned. Indeed, despite the public unease about BST, the FDA approved its use in late 1993 and did so again without requiring that products from BST-treated animals carry any labels indicating it was used.

In fact, when FDA announced its decision not to label, I was at a National Agricultural Biotechnology Commission meeting with hundred of scientists. NABC is a very enlightened group that has long discussed ethical issues in agricultural biotechnology. The group was so horrified at the FDA decision that we all signed a letter protesting the failure to label. Not surprisingly, we were totally ignored.

What sorts of dangers could accompany genetic engineering? There are, of course, dangers associated with traditional selective breeding, but given the multigenerational, inherently slow pace of traditional breeding, we have ample time to watch those dangers emerge and to revise our selection. Even with traditional breeding, unpredictable and unexpected negative results may occur.

There are many instances of this, in fact, even in traditional breeding. One famous example of this concerns corn and grows out of the phenomenon known as *pleiotropy*, which means that one gene and its products control or code for more than a single trait. In this case, breeders were interested in a gene that controlled male sterility in corn, so that one could produce hybrid seeds without detasseling the corn by hand, a task that is very labor-intensive. So the gene was introduced in order to provide genetic detasseling. Unfortunately, the gene was also responsible for increased susceptibility to southern corn blight, a fact no one was aware of. The corn was widely adopted, and in one year a large part of the corn crop was devastated by the disease.

Similarly, when wheat was bred for resistance to a disease called blast, that characteristic was looked at in isolation and was encoded into the organism. The backup gene for general resistance, however, was ignored. As a result, the new organism was very susceptible to all sorts of viruses that, in one generation, mutated sufficiently to devastate the crop.

If these sorts of problems can emerge with traditional breeding, a fortiori they are likely to happen when one inserts a new gene, possibly from a different species. When one inserts a gene into an organism, one cannot anticipate pleiotropic effects, where the gene affects other traits. By the same token, one may have overlooked the need for more than one gene to create the desired phenotypic results. And statistically speaking, just as most mutations are deleterious, most genetic insertions are likely to eventuate in results harmful to the organism.

The second type of danger resulting from fast-lane genetic engineering of animals can be illustrated by reference to food animals. Here the isolated characteristic being engineered into the organism may have unsuspected harmful consequences to humans who consume the resultant animal. Thus, for example, one can imagine genetically engineering faster growth in beef cattle in such a way as to increase certain levels of hormones that, when increased in concentration, turn out to be carcinogens or teratogens (causes of birth defects) for human beings. The deep issue here is that one can genetically engineer traits in animals without a full understanding of the mechanisms involved in phenotypic expression of the traits, with resulting disaster. Ideally, though this is probably not possible either in breeding or creating transgenics, one can mitigate this sort of danger by being extremely cautious in one's engineering until one has at least a reasonable grasp of the physiological mechanisms affected by insertion of a given gene.

Another sort of risk involves genetic engineering that amplifies an already extant problem in agriculture. With the development of industrialized agriculture, we have relentlessly selected for animals and plants on the basis of efficiency and productivity. For example, we grow far fewer strains of wheat than we did seventy years ago, and have far fewer breeds of laying hen. The result is that our crops and livestock are far more susceptible to pathogenic devastation or terrorism. We have put, as it were, all our eggs in one basket. Genetic engineering could accelerate that tendency, with everyone wanting the "superior" plant or animal. In response to the concern that species are being lost, some scientists and the USDA have argued that we can now store all manner of genetic material in a gene library, and in principle resurrect the organisms as needed.

A different area of concern about potential danger associated with transgenic animals arises out of the fact that when one changes animals, one can thereby change the pathogens to which they are host. Such a scenario could arise in a number of different ways. In the first place, if one were genetically engineering the animal in question for disease resistance to a given pathogen, one could thereby unwittingly select for new variations among the natural mutations of that microbe to which the modified animals would not be resistant. This new organism could then be infectious to these animals, to other animals, or to humans. In other words, such genetic engineering could, in essence, become selectional pressure for changing the population of pathogens hosted by the organism.

This is not science fiction. This same sort of thing has occurred, as mentioned earlier, by virtue of widespread use of antibiotics in humans and animals, be it for therapy or for growth promotion in farm animals. Most of us have heard of the battle between pathogen change and antibiotic development. We use a given antibiotic to fight a given disease, say, streptococcal infection. We effectively wipe out the susceptible strain, thereby leaving, as it were, a clear field for another strain not susceptible to that antibiotic. That strain proliferates, while drug companies modify the antibiotic in some manner so that it can engage the newly flourishing pathogen. The latter is effectively destroyed, and the cycle begins again. A similar problem has recently arisen with drug-resistant tuberculosis microbes. The same sort of thing could presumably occur with any form of disease mitigation, including genetic resistance. Once again, though, the possible danger resulting from genetic engineering is no different in principle from what we have already encountered in a different context.

Yet another danger arises from our new ability to genetically engineer animals who model human diseases for which no natural animal model exists. Consider the SCID mouse, a mouse genetically engineered to possess the human immune system, and thus be susceptible to infection with the AIDS virus. Such an animal could infect humans were it to get loose, or the endogenous mouse viruses could interact with the AIDS virus to produce new pathogens with unpredictable characteristics.

One of the most widely discussed set of concerns about genetically engineered animals and plants is environmental and ecological. We have a long history of failing to predict the harmful ecological consequences of putting new animals into an established ecosystem: the importation of a dozen pairs of Australian possums into New Zealand at the turn of the century leading to 60 million such possums preying on indigenous fauna; the importation of rabbits into Australia; the zebra mussel wreaking havoc in the Great Lakes; the tree snake accidentally released in Guam, decimating bird species there; and similar situations with the mongoose in Hawaii and foxes put on islands for raising fur in the Aleutians. There are innumerable such cases involving species whose characteristics were known; how much the more so could genetically engineered animals yield unpredictable environmental damage? We have seen this happen with corn genetically engineered with the Bt (*Bacillus thuringiensis*) gene to kill noxious insects, in fact, killing Monarch butterflies. Salmon genetically engineered to grow bigger and faster than normal salmon could outcompete normal animals in an aquatic ecosystem. To their credit, the Ecological Society and fisheries scientists have addressed such potential disasters in a series of reports, and suggested ways of mitigating such dangers.[4]

The military danger of transgenics is all too clear. Suddenly, we could have the capability of altering pathogens – the AIDS virus,

[4] Tiedje et al., "The Planned Introduction of Genetically Engineered Organisms"; E. M. Hallerman and A. R. Kapuscinski, "Transgenic Fish and Public Policy: Anticipating Environmental Impacts," "Transgenic Fish and Public Policy: Patenting Transgenic Fish," "Transgenic Fish and Public Policy: Regulatory Concerns."

for example – so it is now spread by mosquitoes or is aerosolizable. The dangers arising from military applications of biotechnology demonstrate the naïveté of those who would ban genetic engineering. Biotechnology is not nuclear technology; it does not require great amounts of capital to do transgenic work. Thus, any country – or even any reasonably well-financed terrorist group – could generate biotechnology-based weaponry by hiring the relevant expertise. Even if biotechnology were to be prohibited in the United States, it would not stop – it would simply move somewhere else devoid of restrictions, in the manner of Liberian ship registration, Third World drug testing, and Caribbean banks for drug money. What this would amount to, then, would be society playing ostrich; genetic engineering would not go away. It would simply be out of sight and out of oversight. Thus, any attempt to deny genetic engineering is to lose the charge to keep it under social control and scrutiny, not to eliminate it. That is indeed a major reason why I am writing this chapter. As a society we can no longer afford to engage spurious, sensationalistic issues of the sort I discussed earlier, while the real issues are essentially socially ignored.

Finally, there are socioeconomic dangers associated with genetic engineering. The use of genetically engineered animals or products such as BST would accelerate the replacement of small farms and husbandry agriculture by large corporate entities driven solely by efficiency and productivity and profit, not way of life. Both animals and the environment have been harmed by the advent of a profit-*über-alles* mind-set; it seems evident that genetic engineering would play into this tendency, not subvert it.

As long as public ethical concerns about risk are ignored by scientists, corporations, and regulators (as, for example, we saw in labeling, where the public's right to know and right to choose are cavalierly ignored), public acceptance of biotechnology will necessarily be aborted. Once again, some venue needs to be created where people's ethical concerns are taken seriously and enter into scientists' decisions about creating certain biotechnological modalities, or at least where scientists attempt to persuade the public to change those values. Failure to do this risks rejection of products on which great amounts of money and time have been spent. Such educational dialogue could be undertaken through the establishment of local review committees for proposed biotechnological products.

We have seen that avoidance of dangers does not require a good deal of ethical thought but can be seen as a matter of prudence. The one area of genetic engineering of animals or other biotechnological manipulations that represents a purely ethical issue is the consideration of the effect of such manipulations on the health, welfare, and well-being of those creatures being manipulated: the animals. Much precedent exists for this sort of concern in the Frankenstein myth. A clear theme in the Frankenstein novel is that the monster is reviled, rejected, and rebuked merely in virtue of having been created. Despite its innocence, it suffers. Echoes of this theme appear in almost all the Frankenstein movies and in other horror movies. I particularly recall the Dino de Laurentiis remake of *King Kong*, where the entire theater cheered the "monster" as it was being attacked by the police and the military.

There is every reason to worry about the effect of genetic manipulation on the animals. If one looks at animal use in the twentieth century, it is clear that not only human need but human benefit and economic advantage have tended to trump concerns about animal welfare.

We can readily discern the way in which the development of industrialized confinement agriculture totally destroyed the fair contract and correlative animal welfare consideration historically implicit in animal husbandry. Whereas husbandry agriculture presupposed putting animals into the best possible environment suiting their natures and then augmenting that situation by providing food during famine, water during drought, help in birthing, medical attention, protection from predation, and so on, putting, as it were, square pegs in square holes and round pegs in round holes, industrialized agriculture, through the use of technological "sanders" such as vaccines and antibiotics, allowed us to force square pegs into round holes, animals into environments that increased profit but hurt their well-being.

If one sees this mind-set as dominant in agricultural science, animal welfare takes a solid second seat to efficiency and productivity, if it is indeed given consideration at all. This, in turn, has major implications for the well-being of genetically engineered or otherwise biotechnologically manipulated animals in agriculture. Consider the Beltsville pig, which had the human growth hormone gene inserted into it so that it would grow faster and leaner. (From what we have already discussed, we can see that using a human gene was probably not the wisest thing to do in terms of the first interpretation of the Frankenstein story.)

The desired results were to increase growth rates and weight gain in farm animals, reduce carcass fat, and increase feed efficiency. Although certain of the goals were achieved – in pigs, rate of gain increased by 15 percent, feed efficiency increased by 18 percent, and carcass fat was reduced by 80 percent – unanticipated effects, with significantly negative impact on the animals' well-being, also occurred. Life-shortening pathologic changes, including kidney and liver problems, were noted in many of the animals. The animals also exhibited a wide variety of disease and symptoms, including lethargy, lameness, uncoordinated gait, bulging eyes, thickened skin, gastric ulcers, severe synovitis, degenerative joint disease, heart disease of various kinds, nephritis, and pneumonia. Sexual behavior was anomalous – females were anestrous and boars lacked libido. Other problems included tendencies toward diabetes and compromised immune function.[5] Sheep similarly engineered fared better for the first six months, but then also became unhealthy.[6]

There are certain lessons to be learned from these experiments. In the first place, although similar experiments had been done earlier on mice, mice did not show many of the undesirable side effects. Thus it is difficult to extrapolate in a linear way from species to species when it comes to genetic engineering, even when, on the surface, the same sort of genetic manipulation is being attempted.

Second, as we saw in the discussion on risk, it is impossible to effect simple one-to-one correspondence between gene transfer and the appearance of desired phenotypic traits. Genes may have multiple effects; traits may be under the control of multiple genes. The relevance of this point to welfare is obvious and analogous to a point I made earlier about risk: One should be extremely circumspect in one's engineering until one has a good grasp of the physiological mechanism affected by a gene or set of genes. One good example of the welfare pitfalls is provided by attempts to genetically engineer mice to produce greater amounts of interleukin-4, in order to study certain aspects of the immune system. This, in fact, surprisingly resulted in

[5] Pursel et al., "Genetic Engineering of Livestock."
[6] Fox, *Superpigs and Wondercorn*, p. 117.

these animals experiencing osteoporosis, a disease resulting in bone fragility, clearly a welfare problem.[7]

Another example is provided by a recent attempt to produce cattle genetically engineered for double muscling. Though the calf was born showing no apparent problems, within a month it was unable to stand up on its own, for reasons that are not yet clear.[8] To the researchers' credit the calf was immediately euthanized. Yet another bizarre instance of totally unanticipated welfare problems can be found in the situation where leglessness and craniofacial malformations resulted from the insertion of an apparently totally unrelated gene into mice.[9]

Thus, animal care and use committees should demand that, in pilot research on agricultural animals, a small number of animals be used and that early end points for euthanasia of these animals be established in advance and implemented at the first sign of suffering or problems that lead to suffering, unless such suffering or disease can be medically managed. Such a demand already exists in the laws governing animals used in biomedical research, but, as mentioned earlier, these laws do not apply to food and fiber research. Clearly, this loophole must be closed before genetic engineering of farm animals, even at an experimental level, is allowed to proceed. Currently, the situation is absurd. One of my colleagues has a colony of research sheep. If a ewe of his gives birth to twin lambs, one lamb may go to biomedical research, the sibling to agricultural research. If both go to experiments requiring castration, the biomedical research lamb must have it done under aseptic conditions, with anesthesia and postoperative analgesia. His unfortunate brother, however, can be treated according to accepted agricultural practice, which could include pocketknife castration or even having the testes bitten off.

The larger problem, however, arises not at the level of research but at the level of commercial production. Although it is extremely unlikely that either practical constraints or public opinion would allow the Beltsville pig to be commercially viable (recall that it is infertile), let us imagine that reproduction was not a problem and that the multiple

[7] Lewis et al., "Osteoporosis Induced in Mice by Overproduction of Interleukin-4."
[8] G. Niswender, personal communication.
[9] McNeish et al., "Legless, a Novel Mutation Found in PHT1-1 Transgenic Mice."

problems experienced by the animal did not impair the increased efficiency and productivity accomplished by gene transfer. In other words, imagine that the pig was economically viable, though suffering in a variety of dimensions. Here we would have a situation analogous to some of the more extreme cases in confinement agriculture, where animals may suffer yet be productive or where at least the operation as a whole is productive. There is little doubt that these pigs would be commercially produced, given the nature of the industry's bottom line. In confinement agriculture, profits per animal are small; success is possible only through tiny increases in profit magnified over an operation of significant scale. If one could gain a competitive edge through using these animals, they would enter the production arena. For purposes of economic benefit, the suffering and sickness of the animals would not enter the equation.

To forestall this sort of animal suffering for commercial purposes, I have proposed a regulatory principle I believe should be made into law: the Principle of Conservation of Welfare. What this principle affirms is that *given any proposed genetic engineering of animals, the animals should be no worse off, in terms of suffering, after the new traits are introduced into the genome than the parent stock was prior to the insertion of the new genetic material.* This would forestall the creating of suffering animals for profit or increased "efficiency" but would not stop genetic engineering that benefits the animals and producers, such as increased disease resistance. Most audiences I have described this principle to, including researchers and people in the agricultural industry, would acquiesce to it, as a matter of minimal decency and of staying in accord with what I have described as the "new social ethic for animals." But while this principle may work well for agriculture, it raises a major dilemma in biomedical research.

In a previous chapter, we discussed the emergence of a new ethic for animals focused not on cruelty but on animal suffering resulting from perfectly decent motives. We also indicated that most countries now regulate animal research in accordance with the new ethic as well as an ever-increasing tendency to press forward alternatives to painful animal use. Thus, for example, a January 1998 article in *Lab Animal* indicated that "increasing concern within and without the scientific community over pain and distress in animals has made the production of monoclonal antibodies [MAbs] highly controversial . . . [with] some

European countries having gone as far as banning in vivo production of MAbs using the ascites method."[10] In the United States, pain engendered in laboratory animals (e.g., in studies of tumor growth and disease processes) must be controlled by anesthesia, analgesia, sedation, and early end points, aimed at minimizing suffering. In Britain, an animal suffering uncontrollable pain and distress must be euthanized as soon as the situation is understood.[11] For reasons of controlling pain and suffering, U.S. journals are increasingly unlikely to publish papers using death as an end point, even though the late end point may well provide valuable information. In other words, globally, there is a consensus emerging that not every human benefit is worth any amount of animal suffering (*vide* public rejection of cosmetics companies utilizing safety testing on animals, and the spectacular growth of those companies disavowing such testing).

Everything we have said thus far is patent, undeniable, and clearly points out the profound and worldwide socio-ethical concerns about invasive animal use. The message to researchers is thus clear: Minimize animal suffering; the same message is ever increasingly being sent to animal agriculture, as evidenced in British and European regulations and most notably in the 1988 Swedish law abolishing confinement agriculture. (In the United States the concern with suffering is augmented by an additional concern for environmental enrichment legally mandated for primates and strongly pressed for all species in the most recent NIH *Guide to the Care and Use of Laboratory Animals.*) Both Dr. Tom Wolfle of the Institute for Laboratory Animal Resources at NIH and I have long argued that animals probably suffer more in virtue of the impoverished environments we keep them in for convenience than from the invasive manipulations to which we subject them.

How does this apply to the use of transgenic animals in research? For certain such uses, satisfaction of the demand for control of pain and suffering is precisely analogous to what occurs in research with nontransgenic animals.

Consider, for example, the very first patented transgenic animal, the Harvard mouse that is disposed to the development of tumors. In the words of the patent, this is "an animal whose germ cells and

[10] Shalev, "European and U.S. Regulation of Monoclonal Antibodies."

[11] O'Donoghue, "European Regulation of Animal Experiments."

somatic cells contain an activated oncogene sequence introduced into the animal . . . which increases the probability of the development of neoplasms (particularly malignant tumors) in the animal" (U.S. Patent Number 4,873,191). Minimizing pain and suffering for such an animal is in principle and in fact no different from minimizing pain and suffering in nontransgenic animals in whom tumors are induced by other means: the establishment of end points for euthanasia in terms of tumor size, so that the animal does not suffer, and the judicious use of anesthetics, analgesics, and tranquilizers in the course of operative or other procedures.

By the same token, there is no reason not to apply the other major thrust of the new social ethic to these transgenic animals, namely, the provision of enriched environments and husbandry systems for these animals congenial to their natures, which allow them to actualize their behavior and biological natures. Indeed, the characterization of such environments and systems is a primary purpose of the chapters in the CRC book mentioned earlier.[12] Thus, in the case of transgenic mice, one should look to the recommendations in the general literature on care of mice; for example, a British article described a caging system for rodents that is meant to accommodate their behavioral needs.[13] Thus, the vast majority of transgenic animals developed thus far raises no additional welfare issues beyond those concerning nontransgenic laboratory animals.

Indeed, those welfare issues that are raised dramatically by transgenic animals are also continuous with analogous nontransgenic cases. I am referring to the creation and maintenance of seriously genetically defective animals developed and propagated to model some human genetic disease. This was traditionally accomplished through identification of adventitious mutations and selective breeding. Transgenic technology allows for accomplishing the same goal far more quickly and in a far wider range of areas. Thus, one can, in principle, essentially replicate any human genetic disease in animals. And therein lies the major ethical concern growing out of transgenic technology in the research area. It is a true dilemma, because there are strong moral pulls on both sides of the issue.

[12] Rollin and Kesel, *Experimental Animal in Biomedical Research,* vols. 1 and 2.
[13] Sharmann, "Improved Housing of Mice, Rats, and Guinea Pigs."

A chapter in a book devoted to transgenic animals helps to focus the concern:

There are over 3,000 known genetic diseases. The medical costs as well as the social and emotional costs of genetic disease are enormous. Monogenic diseases account for 10% of all admissions to pediatric hospitals in North America . . . and 8.5% of all pediatric deaths. . . . They affect 1% of all live born infants . . . and they cause 7% of stillbirths and neonatal deaths. . . . Those survivors with genetic diseases frequently have significant physical, developmental or social impairment. . . . At present, medical intervention provides complete relief in only about 12% of Mendelian single-gene diseases; in nearly half of all cases, attempts at therapy provide no help at all.[14]

This is the context in which one needs to think about the animal welfare issues growing out of a dilemma associated with transgenic animals used in biomedical research. On the one hand, it is clear that researchers will embrace the creation of animal models of human genetic disease as soon as it is technically feasible to do so. Such models, which introduce the defective human genetic machinery into the animal genome, appear to researchers to provide convenient, inexpensive, and most important, "high-fidelity" models for the study of the gruesome panoply of human genetic diseases outlined in the over three thousand pages of text comprising the sixth edition of the standard work on genetic disease, *The Metabolic Basis of Inherited Disease.*[15] Such high-fidelity models may occasionally reduce the numbers of animals used in research, a major consideration for animal welfare, but are more likely to increase the numbers as more researchers engage in hitherto impossible animal research. On the other hand, the creation of such animals can generate inestimable amounts of pain and suffering for these animals, since genetic diseases, as mentioned above, often involve symptoms of great severity. The obvious question then becomes the following: Given that such animals will surely be developed wherever possible for the full range of human genetic disease, how can one assure that vast numbers of these animals do not live lives of constant pain, suffering, and distress? Further, given the emerging ethic we outlined above, control of pain and suffering is a sine qua non for continued social acceptance of animal research.

[14] E. M. Karson, "Principles of Gene Transfer and the Treatment of Disease."
[15] Scriver et al., *Metabolic Basis of Inherited Disease.*

Merely citing the potential human benefit that can emerge from long-term studies of suffering animals created to model human disease won't do – we have already seen that society has rejected that claim about death or advanced stages of a disease as end points. In today's moral ethos, it is simply not the case that any possible human benefits will outweigh any amount of animal suffering. If a genetic disease is rare, affects only small numbers of people, and can be prevented by genetic screening and what Kelley and Wyngaarden call in reference to Lesch-Nyhan's syndrome "therapeutic abortion,"[16] it is not clear that society will accept the long-term suffering of vast numbers of animals as a price for research on the disease. More and more, a cost-benefit mind-set is emerging vis-à-vis animal use in science, just as it is legally mandated for research on humans – though it is by no means clear how one rationally weighs animal cost against human benefit.

To flesh out our discussion with a real example, let us examine the very first attempt to produce an animal "model" for human genetic disease by transgenic means, that is, the development, by embryonic stem cell technology, of a mouse that was to replicate Lesch-Nyhan's disease, or hypoxanthine-guanine phosphororibosyl-transferase (HRPT) deficiency.[17] Lesch-Nyhan's disease is a particularly horrible genetic disease, leading to a "devastating and untreatable neurologic and behavioral disorder."[18] Patients rarely live beyond their third decade and suffer from spasticity, mental retardation, and choreoathetosis. The most unforgettable and striking aspect of the disease, however, is an irresistible compulsion to self-mutilate, usually manifesting itself as biting fingers and lips. The following clinical description conveys the terrible nature of the disease:

> The most striking neurological feature of the Lesch-Nyhan syndrome is compulsive self-destructive behavior. Between 2 and 16 years of age, affected children begin to bite their fingers, lips and buccal mucosa. This compulsion for self-mutilation becomes so extreme that it may be necessary to keep the

[16] Kelley and Wyngaarden, "Clinical Syndromes Associated with Hypoxanthine-Guanine Phosphororibosyltransferase Deficiency."

[17] Hooper et al., "HPRT-Deficient (Lesch-Nyhan) Mouse Embryos Derived from Germline Colonization by Cultured Cells"; Kuehn et al., "A Potential Model for Lesch-Nyhan Syndrome through Introduction of HPRT Mutations into Mice."

[18] Kelley and Wyngaarden, "Clinical Syndromes."

elbows in extension with splints, or to wrap the hand with gauze or restrain them in some other manner. In several patients mutilation of lips could only be controlled by extraction of teeth.

The compulsive urge to inflict painful wounds appears to grip the patient irresistibly. Often he will be content until one begins to remove an arm splint. At this point a communicative patient will plead that the restraints be left alone. If one continues in freeing the arm, the patient will become extremely agitated and upset. Finally, when completely unrestrained, he will begin to put the fingers into his mouth. An older patient will plead for help, and if one then takes hold of the arm that has previously been freed, the patient will show obvious relief. The apparent urge to bite fingers is often not symmetrical. In many patients it is possible to leave one arm unrestrained without concern, even though freeing the other would result in an immediate attempt at self-mutilation.

These patients also attempt to injure themselves in other ways, by hitting their heads against inanimate objects or by placing their extremities in dangerous places, such as between spokes of a wheelchair. If the hands are unrestrained, their mutilation becomes the patient's main concern, and effort to inflict injury in some other manner seems to be sublimated.[19]

At present, "there is no effective therapy for the neurologic complications for the Lesch-Nyhan's syndrome."[20] Thus Kelley and Wyngaarden, in their chapter on HPRT deficiency diseases, boldly suggest as alluded to earlier, "the preferred form of therapy for complete HPRT deficiency [Lesch-Nyhan's syndrome] at the present time is prevention," that is, "therapeutic abortion."[21] This disease is so dramatic that I predicted in 1976 that it would probably be the first disease for which genetic researchers would attempt to create a model by genetic engineering. Researchers have, furthermore, sought animal models for this syndrome for decades and have in fact created rats and monkeys that will self-mutilate by administration of caffeine drugs.[22] It is thus not surprising that it was the first disease genetically engineered by embryonic stem cell technology. But to the surprise of the researchers, these animals, although they lacked the HPRT enzyme, were phenotypically normal and displayed none of the metabolic or neurologic symptoms characteristic of the disease in humans. The reason for the failure

[19] Ibid.
[20] Stout and Caskey, "Hypoxanthine Phosphororibosyltransferase Deficiency."
[21] Kelley and Wyngaarden, "Clinical Syndromes."
[22] Boyd et al., "Chronic Oral Toxicity of Caffeine."

of this transgenic "model" has been suggested to be the presence of a backup gene for xanthine metabolism in mice,[23] though other research has cast doubt on this notion.[24] Though an asymptomatic mouse is still a useful research animal, for example, to begin to test gene therapy,[25] clearly a symptomatic animal would, as a matter of logic, represent a higher fidelity model of human disease, assuming the relevant metabolic pathways have been replicated. Presumably, too, it is simply a matter of time before researchers succeed in producing symptomatic animals – I have been told in confidence of one lab that seems to be close to doing so, albeit in a different species of animal. One may perhaps need to move up to monkeys to achieve replication of the behavioral aberrations.

The practical moral question that arises, then, is clear: Given that researchers will certainly generate such animals as quickly as they are able to do so, how can one assure that the animals live lives that are not characterized by the same pain and distress that they are created to model, especially since such animals will surely be used for long-term studies of the development of genetic diseases? Or should such animal creation be forbidden by legislation, the way we forbid multiple use of animals in unrelated surgical protocols in the United States or the British forbid learned helplessness studies?

There is, admittedly, no absolute or direct proof that U.S. society at least will reject the creation of such animals. The proof is indirect, based on Gaskell's survey in Europe that morally rejected genetic engineering of animal models of disease and based on the incompatibility of creating such animals with the direction in which worldwide attitudes and laws regarding animal research are moving. At the very least, however, it would be prudentially unwise for the research community to forge ahead cavalierly with the creation of such animals for long-term use. For if U.S. attitudes are analogous to European ones, such proliferation of suffering animals could well evoke significant legislative restriction or even banning of any transgenic animal work,

[23] Redhead et al., "Mice with Adenine Phosphororibosyltransferase Deficiency Develop Fatal 2,8-Dihydroxyadenine Lithiasis."

[24] Engle et al., "HPRT-APRT-Deficient Mice Are Not a Model for Lesch-Nyhan Syndrome."

[25] Jinnah et al., "Dopamine Deficiency in a Genetic Mouse Model of Lesch-Nyhan Disease."

including the sort of work where life-long suffering can be avoided by early end points, anesthesia, and so on.

In preparing a paper on this issue I felt the need to test common sense's reaction to the dilemma we have described above with regard to Lesch-Nyhan's disease. What I am about to report is by no means scientifically sound or statistically significant. Nonetheless, I think it is at least an indicator of changing attitudes researchers must reckon with, in the same way that talking to friends and acquaintances about former President Clinton's sexual hijinks gave one an indication that while people didn't like it, they seemed to separate that behavior from his ability to govern.

Specifically, I approached some forty students in an agricultural ethics course I teach for the CSU College of Agriculture. Over 95 percent of the students are involved in animal agriculture, generally cattle. Thus, if there is a bias in this sample, it is surely toward animal use. I explained the dilemma we are discussing to them in a straight-forward way, using Lesch-Nyhan's as an example, and asked them for their view of resolving it. Much to my amazement, there was virtual unanimity in the group against producing such models. Some of the opinions expressed in support of their position were as follows:

1. Nature makes mistakes; we should not try to fix them all.
2. If only a small number of people are affected, it would be wrong to create a large number of suffering animals in order to study it.
3. Lesch-Nyhan's should be dealt with by genetic screening and abortion.
4. Not every human benefit – even cure of disease – justifies limitless amounts of animal suffering.
5. All acknowledged that they might feel differently if a close member of their family was suffering from the disease.
6. Few had problems with using large numbers of animals to study anything, as long as the animals don't suffer.
7. Some felt that results obtained for the human genome project might yield results obviating the need to create animal models of such diseases.

At any rate, the issue of genetically engineering chronically suffering animals is clearly a serious one, and one that any conscientious animal researcher ought at least to perceive as a dilemma, with strong ethical pulls in opposite directions. It is an issue I first addressed in 1985 and have continued to address in numerous other venues during the ensuing years at conferences and in papers. The tack I have taken was to throw the dilemma out to researchers and veterinarians and ask for their ideas, in hope of finding a "middle way" to pass between the horns of the dilemma. I even phoned a high-ranking colleague at NIH and asked whether they were addressing the high potential for pain and suffering in animal models of genetic disease. Unfortunately, I never received any creative suggestions, though virtually all audiences were sensitive to the moral problem raised and also to the pragmatic dangers to public support of animal research were such research to move blithely forward while ignoring the pain and suffering issue. Although numerous veterinarians and some scientists responded by forthrightly offering, "Perhaps we shouldn't create such animals," and no one responded by saying, "To hell with the animals – we're saving people," few doubted that researchers would create such "models" as soon as they were technically capable of doing so.

Seeing that no one was pursuing a viable solution, I felt a moral imperative to seek one myself. And during the 1990s, my strategy was to think in terms of long-term anesthesia modalities to ensure that transgenic animals modeling diseases such as Lesch-Nyhan's would not suffer. Knowing that no one had anesthetized an animal for more than six weeks, and then only with Herculean and expensive effort, and further knowing that such animals would be used for long-term studies in order to uncover the development of disease, I, together with dedicated and morally concerned colleagues at CSU, ruled out long-term anesthesia. We then considered creating coma in the animals for purposes of ablating awareness of pain and distress but found that this is technically nonfeasible in mice, both because it is virtually impossible to intubate them and because researchers would probably question the value of such a model, since the dramatic symptoms of Lesch-Nyhan's, at least, may well involve the cerebral cortex. Similar difficulties accompanied the idea of rendering the animals decerebrate either surgically or genetically. In any case, the problems of maintaining vegetatively alive, unconscious animals for long periods are staggering: intraperitoneal (IP) feeding, bladder expressing, keeping lungs clear, drying of eyes and respiratory membranes, and so on. No additional suggestions were forthcoming, and though I continued to raise the problem, I grew increasingly pessimistic at finding a solution.

Being invited to an American College of Laboratory Animal Medicine symposium to speak on the issue served to, as Kant once said, shake me out of my dogmatic slumber. I resolved to find a way to deal with the dilemma in a viable fashion, not sing the same old song.

Society does not wish to see the creation of chronically suffering animals. Human medical researchers do not wish to abandon the potential vehicles for study of human genetic disease that transgenic animal models provide. One easy solution, of course, is to say, as some members of my audiences have done, "Don't create such animals." But I doubt that researchers in the area would agree with such a response, both for reasons of principle concerning closing down of whole avenues of research and also because of the potential benefits for human health. On the other hand, such researchers must (or should) surely be aware that they are bound to respect socio-ethical concerns, since public money often pays for research and also because public moral rejection of their work could close them down. To adopt a stance saying that only by being allowed to create life-long suffering animals can we help humans is to invite being shut down in that area of transgenic work, even as research into cloning of humans has been shut down by virtue of public moral revulsion. Indeed, a cavalier stance about pain and suffering in genetically engineered animal models could seriously negatively affect all transgenic animal research, even that not involving pain and suffering, for the public could equate all transgenic animals created for research with animals suffering uncontrollable, life-long pain. Further, even if the European public rejected genetically engineered animal models for some reason other than suffering, that is, some bad ethical reason, public knowledge of the creation of animals experiencing life-long suffering and pain could only underscore and further solidify that rejection. So the issue must be dealt with by the research community both for ethical reasons and reasons of preservation of autonomy. How can this be accomplished?

To answer this question, we must recall what we said about the nature of social and professional ethics. Every society, to assure social order, must articulate a *social consensus ethic* governing matters deemed essential. This is usually "written large" in Plato's phrase, in the legal system. Thus we morally and legally disallow rape, bank robbery, murder, fraud, and so on. Matters with moral import not affecting the social order are left, generally, to one's personal ethic – for example, which religion one believes in, to whom and whether one gives charity, what one reads, and so on. Things move in and out of the social ethic as society changes – thus, control of nonviolent sexual behavior was surrendered to the personal ethic from the social ethics, beginning with the 1960s' "sexual revolution," yet at roughly the same time hiring and firing or selling and renting of property were appropriated by the social ethic from the personal because leaving it to the personal was perceived as generating unfairness and injustice via discrimination. In general, the social ethic appropriates matters when leaving them to personal ethics is perceived as leading to morally unacceptable behavior. During the past twenty-five years, increasing amounts of animal treatment have been moved into the social ethic for these reasons, for example, laws regulating research, agriculture or animal shows.

Professional ethics stands midway between social and personal and applies to subgroups of society engaged in important but highly specialized activities not well understood by society in general and requiring special expertise and special privileges. Thus, veterinary or human medicine requires specialized knowledge and enjoys special privileges – for example, dispensing pharmaceuticals or performing surgery. Society says, in essence, to professions: "You regulate yourselves the way we would regulate you if we understood what you were doing well enough to do so." Failure to meet this demand can result in ill-informed people regulating the profession in question without understanding.

Animal researchers are, of course, such a professional subgroup. When they failed to control pain and suffering or provide good animal care, as we saw above, U.S. society moved in 1985 to regulate animal research despite protestations that such regulation would endanger human health. (In fact, it did no such thing, and almost certainly led to better research.) If they fail to consider – and implement – the socio-ethical requirement of not creating animals who suffer greatly for long periods of time, society will almost certainly move to regulate such activities, even if researchers again protest that such regulation endangers human health.

I would therefore argue that the research community needs to press forward, on their own, the following sort of regulatory requirement, to be binding on all transgenic animal research and enforced by animal care and use committees: *No one may create a transgenic model of human genetic disease until he or she has provided a method for assuring that the animals do not suffer uncontrollable long-term or lifetime pain.* (A precedent for this already exists in animal care and use committees generally not allowing animals used for disease research to progress to death as an end point.) Such a principle would put the burden for the control of suffering where it should be: on the researcher and the ACUCs. Just as current law requires the control of postsurgical pain in animals used in research but does not specify how this is to be done, so the above principle precludes the making of long-term suffering transgenic animals and, rather than simply prohibiting the creation of such animals, instead places the burden for controlling pain and suffering on those who propose such "models."

It is possible, of course, that researchers may be unable – even as I was unable – to come up with modalities that satisfy the above principle, thus in effect making the principle essentially equivalent to a prohibition regarding creation of such animals, but there is nonetheless a significant difference between simple prohibition and our proposed principle. The difference is that the moral burden of social concern is placed on researchers, rather than society simply setting the precedent of dictating by fiat on matters it doesn't understand. There is a huge conceptual difference between saying "You may not pursue a line of research" and saying "You may pursue it if you control pain and suffering." The latter is an extension of established social ethics; the former sets the precedent of significant social intrusion into scientific autonomy without allowing science to demonstrate that it can pursue the research in a manner acceptable to social ethics.

Were such a principle to be adopted in the United States, it certainly goes beyond current law. At the moment, a researcher must control pain and suffering or provide a justification for not doing so. Our principle eliminates the "escape hatch," in essence making control of pain and suffering a nonnegotiable requirement. Why should researchers acquiesce to this, when in fact under current law they could probably convince at least some animal care and use committees of the value of creating such animals, even when one can't control the pain and suffering? The pragmatic answer I would tender is twofold. First, I seriously doubt, as Gaskell's report indicates, that society will tolerate the creation of such animals. Were news media to inform the general public of the deliberate creation of life-long suffering animals, especially if photographs or video were forthcoming, few can doubt the strong visceral reaction this would engender. Just as the University of Pennsylvania head injury lab films in essence forced the passage of the 1985 laboratory animal laws, such footage could be very well lead to strong regulatory action that could be more restrictive than the principle we formulated. Second, failure to address this issue would surely awaken and give succor to all the social, knee-jerk "Frankenstein syndrome" reactions to genetic engineering and could further alienate the public from biotechnology in general.

But in the end, the deeper answer is a purely moral one. Twenty or more years ago, when animal researchers were typically trained under the ideology affirming that science was "value- and ethics-free" and that felt pain in animals was not scientifically knowable, they could distance themselves more easily from the suffering they created. As this ideology has crumbled and been replaced by a more reasonable view, scientists have inevitably developed closer kinship with social morality, since it is now more difficult to distance one's role in science from one's ordinary common sense and morality. And few things are harder for a morally reflective individual to tolerate than dooming something sentient to a life of pain and suffering that cannot be alleviated. Thus the principle we have argued for is simply in the end a corollary of common decency.

Biotechnology raises another vexatious problem for animal use in research ethics. We saw earlier that the mainstay of assuring ethical research was the animal care and use committee. That, in turn, depends on prospective review of protocols. But increasing amounts of biomedical research now involve "let's see what happens if" research involving the insertion and ablation of genes and/or fragments. As we mentioned earlier, the effects of such manipulations are unpredictable. If that is the case, how can a committee assure that proper anesthesia, analgesia, end points, and other modalities to control pain are done in a timely fashion, if one can't anticipate such eventualities?

The only possible answer is to develop sophistication in animal technicians at detecting pain, suffering, and distress. Failure to address this new problem could result in unanswered animal suffering and loss of public confidence in current mechanisms.

8

Biotechnology and Ethics III

Cloning, Xenotransplantation, and Stem Cells

The template we have developed for discussing genetic engineering, wherein we distinguished three possible types of issues – intrinsic wrongness, bad results, and harm to the object of biotechnological ministrations – can be used to analyze all other areas of biotechnology as well.

As remarked earlier, there is no area of biotechnology that better illustrates our Gresham's law for ethics than cloning. Indeed, the failure of the scientific community to prepare the public in advance to discuss rationally the ethical issues involved in cloning, or even to discuss such issues when the birth of Dolly the cloned sheep was in fact announced, resulted in the proliferation of bad ethics and in those ideas becoming ensconced in the public discussion.

Although the cloning of Dolly was enthusiastically acclaimed by most biologists, such was not the case in society in general. Theologians predictably condemned humans usurping the role of God and called for a ban on such research. But the general public too expressed fear, revulsion, and "ethical concern," prompting the British government to cut off additional funding to related research. According to a CNN/*Time* magazine survey of 1,005 adult Americans released one week after Dolly's birth announcement, "most Americans think it is morally unacceptable to clone either animals or humans, and that new cloning techniques will create more problems than they solve."[1]

[1] CNN/*Time* poll, "Most Americans Say Cloning Is Wrong."

Some 54 percent said they would not eat cloned animals, while more than half said they would eat cloned plants, and two-thirds said that the federal government should regulate cloning of animals. Even greater apprehension greeted the possibility of cloning humans. Some 69 percent said that they are "scared" by the prospect of cloning humans, and 89 percent said that such cloning would be "morally unacceptable." Some 75 percent said that cloning is "against God's will," and 29 percent were sufficiently troubled by the new technique for reproducing life that they would be willing to participate in a demonstration against cloning humans. Only 7 percent said that they would allow themselves to be cloned if given the chance.

Media coverage of Dolly was considerable, and the story was front-page news for days. As one who writes in bioethics, I received numerous phone calls from newspapers and broadcasters in a single day, as well as numerous calls and visits from lay people – and scientists – worried about the "ethics" and "ethical issues" occasioned by cloning.

Despite the pervasive outcry about the "ethical issues" and "moral unacceptability" of cloning, there was little rational articulation or clarification of what these issues in fact are, or why cloning is morally unacceptable. Instead, as just mentioned, one heard a great deal about cloning "violating God's will," being "against nature," or, attesting to the power of margarine advertising to shape Western thought, it "not being nice to fool Mother Nature." Further, much of the concern was based on factual misinformation. For example, Pope John Paul II affirmed that cloning was wrong because all creatures had the right to a natural birth, thereby both forgetting that Dolly did have a natural birth and that in any case he was begging the question had she not had one. Others, apparently unaware that Dolly had been born some seven months before her birth was announced, assumed that cloning gave rise to fully formed adults. Still others assumed in their critique that, through cloning, one could produce indefinite numbers of *literally identical* individuals, not just *genetically identical* individuals – as we shall see, the two concepts are very different.

Predictably, bad ethics reached a cacophony, at the (at this point speculative) prospect of cloning human beings. Although the scientist who cloned Dolly had not said a word about the ethics of animal cloning, he did immediately weigh in on the side of the view that cloning humans was morally unacceptable. Religious leaders entered the fray with a vengeance; the public too invoked religious arguments.

The first point is to stress is that religious or theological pronouncements or beliefs, however ubiquitous, even those extending to 100 percent of the population, do not necessarily constitute a moral argument. It is, of course, often the case that moral injunctions are couched in religious terms – *vide* at least some of the Ten Commandments – but the moral content must be in principle conceptually separable from the religious for it to constitute a truly moral claim. Similarly, much of our social ethic is doubtless historically derived from religious sources, but we do not validate our ethics by appeal to religion in a secular society. On the other hand, religious traditions have indeed given much thought to how humans ought morally to live, and much of that thinking may be viable even outside the theological tradition in which it is embedded. For example, the Thomistic moral-psychological point that society should prohibit overt cruelty to animals even though animals lack immortal souls, and thus are not direct objects of moral concern, since people who are allowed to abuse animals are psychologically disposed to abusing people, can be restated independently of the theological assumption about animals' lacking soul. So restated, it has received succor from twentieth-century socio-psychological research linking early animal abuse with subsequent abuse of humans.

Merely because a practice violates a religious tradition does not render that practice immoral for those of us outside that tradition, which in terms of social ethics in a secular society is all of us. For example, many women's styles violate the moral precepts of orthodox Judaism or Islam, yet we do not consider women so attired as socially immoral. To be sure, religious traditions may work to affect our social morality and may even do so, as in Sunday blue laws, but the results must putatively and theoretically stand as moral principles independently of their religious origins. In those cases where they do not, and instead bespeak inordinate political power of a religious tradition, we are rightly suspicious and fear a loss of separation of church and state.

There is nothing wrong with looking to pronouncements from religious leaders as a heuristic device when one is attempting to unpack the moral implications of some new technology such as cloning – many religious leaders may be well-trained in identification of moral issues. A problem arises only if we assume that anything declared to be a moral issue by such leaders automatically becomes one, in a performative way. As we look to such pronouncements, we must be prepared to recast them in strictly secular moral terms and, if we cannot, to reject them as not morally significant. The Pope, for example, may be infallible in matters of faith and Catholic morals, but certainly not in secular ethics.

Having said all this, what sorts of religious ethical concerns emerged at the prospect of human cloning? Not surprisingly, religious leaders, even within a given community of faith, did not speak with one voice. Even for strict fundamentalists possessed of ancient sacred texts, there is little likelihood of univocal readings of those texts as applied to current technological innovations. Thus, while some Christians have condemned genetic engineering, others have said that, like other works of man, it is acceptable as continuing human "co-creation" of nature in partnership with God.

Surprisingly, the claims about cloning advanced by religious leaders yielded significantly divergent pronouncements and arguments, some of which can be distilled down to sensible moral claims, while others cannot. In this portion of our discussion, we will focus on those that cannot, that is, on the claims that cloning is intrinsically or "just" wrong.

A particularly valuable source of religious and cultural perspectives on cloning was a newsletter, *Reflections*, which garnered comments from representatives of almost every major faith tradition in America.[2] One argument, from Stanley Harakas, a Greek Orthodox theologian, affirmed the following:

> Whatever motivations and intentions there might be to take this immoral step, I can think of none that would escape the charge of manufacturing a human being for the purpose of exploiting him or her in a way that depersonalizes the human clone.... Further, in itself, cloning would violate practically every sacramental dimension of marriage, family life, physical and spiritual nurture, and the integrity and dignity of the human person.[3]

Insofar as one can extract an argument from this claim, it appears to be that there is no possible way in which a human derived by cloning

[2] Woolfrey and Campbell, *Reflections.*

[3] Ibid., p. 3.

could enjoy full humanity by virtue of the fact that the motivation for so creating a person would of necessity be exploitative.

This is clearly question-begging. Below on in the discussion I cite an example where it would seem perfectly reasonable to clone a child for decent, very humane reasons. Further, even if a child is created for exploitative reasons, it is difficult to see why such a child would not enjoy full personhood. Many humans have children for exploitative reasons – for example, to care for them later in life or to have someone to shape – yet the child's personhood does not depend on the parent's intentions. The claim clearly commits a version of the genetic fallacy, confusing the origin of a child with the subsequent moral status. A human created by cloning is still a human, even as one can love and respect an adopted child as much as a biological one.

Although I have always found the phrase "human dignity" to be more a matter of moral rhetoric than content, it is difficult to see why one's dignity should depend on one's origin rather than on one's nature. Nonetheless, the claim that clonal derivation deprives humans of dignity is a common reason for rejecting it out of hand as just wrong.[4] Some notion of the elusiveness of this concept can be garnered from the testimony of Catholic Theologian Albert Moraczewski before the National Bioethics Advisory Commission. (NBAC): "In the cloning of humans there is an affront to human dignity. . . . Yet in no way is the human dignity of the [cloned] person diminished."[5] One is tempted to invoke the old logical positivist category of "nonsense" here.

One perennial theme that emerges in attacks on the intrinsic wrongness of cloning and genetic engineering is the "playing God" claim – that such activities involve humans aspiring to decision making that ought to be reserved to the deity. In the National Bioethics Advisory Commission Report, this claim is "unpacked" in the following way:

This slogan is usually invoked as a moral stop sign to some scientific research or medical practice on the basis of one or more of the following distinctions between human beings and God:

- Human beings should not probe the fundamental secrets or mysteries of life, which belong to God.

[4] Ibid., p. 4; National Bioethics Advisory Commission (NBAC), *Cloning Human Beings*, pp. 42–49.
[5] Ibid., pp. 44–45.

- Human beings lack the authority to make certain decisions about the beginning or ending of life. Such decisions are reserved to divine sovereignty.
- Human beings are fallible and also tend to evaluate actions according to their narrow, partial, and frequently self-interested perspectives.
- Human beings do not have the knowledge, especially knowledge of outcomes of actions attributed to divine omniscience.
- Human beings do not have the power to control the outcomes of actions or processes that is a mark of divine omnipotence.

The warning against playing God serves to remind human beings of their finiteness and fallibility. By not recognizing appropriate limits and constraints on scientific aspirations, humans reenact the Promethean assertion of pride or hubris. In the initial theological discussions of cloning humans, Paul Ramsey summarized his objections by asserting: "Men ought not to play God before they can learn to be men, and after they have learned to be men, they will not play God."[6]

While such an account presumably makes sense within the theological context or universe of discourse of Judaeo-Christianity but not, notably, of Hinduism or Buddism, it is difficult to extract secular moral sense from it, save by seeing it as an admonishment against human "arrogance." As the NBAC summary puts it: "If making people in your laboratory isn't playing God, the phrase has no meaning."[7]

There is a serious point to such warnings, but it is, first of all, not restricted or special to cloning or biotechnology, and, second, it does not justify the intrinsic wrongness of cloning but only stresses the possibility of unanticipated risks that may emerge from it. Let us pause briefly to examine these points.

The theme of humans sawing off limbs on which they are seated, painting themselves resolutely into corners, or being left up the creek without a paddle is an ancient one. The aforementioned chutzpah stories of the Tower of Babel, Daedalus and Icarus, the Golem, Frankenstein, and the Sorcerer's Apprentice all warn of excessive optimism by humans in deploying new knowledge or *techne*. "Oops" should be the logo for humanity.

A *locus classicus* for the discussion of this phenomenon is David Ehrenfeld's *The Arrogance of Humanism*, wherein Ehrenfeld relates

[6] NBAC, *Cloning Human Beings*, "Summary of Arguments against Cloning," p. 2.
[7] Ibid.

countless examples of humans marching resolutely off cliffs, from the unanticipated consequences of building the Aswan Dam to various other ecological disasters we have perpetrated.[8] After living through sundry Ebola spills, *Challenger* and Chernobyl disasters, "killer bee" escapes, and iatrogenic diseases, society is far less likely to believe "Trust me, I'm a scientist." And things are no different in the area of biotechnology. In fact, I have argued elsewhere that only when society in general is consulted on biotechnology, through some such vehicle as local review of projects, will society feel a stake in it. And this is as it should be in a democratic society.

Yet there is nothing especially arrogant about human cloning – indeed, the scientific community has itself been relatively circumspect in not pressing it forward. In my view, the creation of herbicide-resistant crops without public discussion is far more "arrogant," as is basing admission into medical schools in large part on a few hours of "scientific" exams. The key point, in any case, is that the whole concern about arrogance associated with cloning is a *consequentialist* one, based on dangerous unanticipated results, not on inherent wrongness.

Enough has been said, I hope, to lay bare the point that arguments about the intrinsic wrongness of cloning are

(1) meaningless or
(2) may have meaning within a given religious tradition but are not subject to secular translation or
(3) *evolve* into points about the likely and significant dangers that cloning must inevitably engender.

(1) and (2) are clearly irrelevant to social decision making, so we must turn to the second aspect of the Frankenstein story as it applies to human cloning, namely, the manifest dangers and risks it will likely occasion.

Are there any risks associated with cloning animals? The most significant risk seems to me to arise from the potential use of cloning to narrow the gene pool of animals, particularly in agriculture. With the advent in the mid-twentieth century of an agriculture based in a business model, and emphasizing efficiency and productivity rather than husbandry and way of life as supreme values for agriculture,

[8] Ehrenfeld, *Arrogance of Humanism.*

it is now evident that sustainability has suffered at the hands of productivity – we have sacrificed water quality to pesticides, herbicides, and disposal of animal wastes; soil quality to sodbusting and high tillage; energy resources to production; air quality to efficiency (as measured by profit); small farms and rural communities to industrialized agriculture. What is less recognized but equally significant is that we have also sacrificed genetic diversity on the same altar.

A lecture I once attended by one of the founders of battery cage systems for laying hens provides an excellent example of how this works. He explained that, with the rise of highly mechanized egg factories, the only trait valued in chickens was high production, that is, numbers of sellable eggs laid. Laying-hen genetics focused with great skill and success on productivity. Inevitably, the production race was won by a few strains of chicken, with other traits deemed of lesser significance. Given the efficiency of artificial selection and rapid generational turnover in chickens, the laying chicken genome grew significantly narrower. Thus today's laying hens are far more genetically uniform than those extant in the 1930s. In fact, said the speaker, such selection has so significantly narrowed the gene pool that, had he known this consequence, he would never have developed these systems.

Why not? Because the narrowing of the gene pool in essence involves (pardon the pun) putting all our eggs in one basket and reduces the potential of the species to respond to challenges from the environment. Given the advent of a new pathogen or other dramatic changes, the population of laying hens could all be decimated or even permanently destroyed because of our inability to manage the pathogen. The presence of genetically diverse chickens, on the other hand, increases the likelihood of finding some strains of animals able to weather the challenge.

Cloning will almost inevitably augment modern agriculture's tendency toward *monoculture,* that is, cultivation and propagation only of genomes that promise or deliver maximal productivity at the expense of genetic diversity. Thus, for example, given a highly productive dairy cow, there will be a strong and inevitable tendency for dairy farmers to clone her and stock one's herd with such clones. And cloning for such a purpose could surely accelerate monoculture in all branches of animal agriculture. Cloning could also accelerate our faddish tendency to proliferate what we think are exemplary animals, rather than what we might really need. For example, very high production milk cows for which we have selected have very short productive lives; very lean pigs are highly responsive to stress; and so on.

At the moment, agriculture's only safety net against ravaged monocultures are hobby fanciers and breeders. Although egg producers disdain all but productive strains, chicken fanciers, hobby breeders, and showmen perpetuate many exotic strains of chickens. Given a catastrophe, it would surely be difficult to diversify commercial flocks beginning with hobby animals as seed stock, but at least genetic diversity will have been preserved.

One suggestion that has been made to prevent loss of genetic diversity is the establishment of a "gene library," wherein one could preserve the DNA of animals no longer commercially useful, and also that of endangered species to prevent their extinction. A major problem with this concept is that the genome in question is, as it were, temporally frozen. Whereas chickens in the world would be adapting to changing conditions, albeit probably slowly and imperceptibly, the stored genome would not. Thus the genome stored in 2005 might not have kept up with environmental changes that have occurred by the time we need to resurrect these chickens in 2905. A second, minor problem arises from the probable slowness of the response time required to resurrect the genome so that it is functionally operative in the requisite environment.

There is also a deeper problem with gene libraries. Such repositories for DNA might well create a mind-set that is cavalier about the disappearance of phenotypic instances of endangered organisms. If people believe that we have the species in a state of, as it were, resurrectable suspended animation, they might well cease to worry about extinction. This is problematic on a variety of levels. First, there is a huge aesthetic (and metaphysical and perhaps ethical) gap between preserving white tigers on the tundra in Siberia and preserving them in zoos, let alone in blueprint form in the test tubes. Furthermore, the tiger does not express its full *telos*, its tigerness, independently of the environment to which it is adapted. To preserve the tiger while allowing its *umwelt* to disappear is to create the biological (and moral) equivalent of stateless, displaced persons. The prospect of white tigers with no place to be in the world but casinos in Las Vegas, wild animal parks, or zoos is not an acceptable solution to the problem of endangered species. Indeed, one could argue that such a solution is little better than extinction, since one has de facto extinguished the tiger as a form of life, though not as a life form.

Let us turn to our next category of issues: Are there any risks associated with cloning people? (For the sake of argument, for the moment our discussion assumes that the cloning process does not harm or cause disease problems in people so derived. I discuss harm to the clone below.)

The standard response, thanks to *The Boys from Brazil*, is the specter of cloning dozens, scores, hundreds, or thousands of Hitlers, all identically evil, demented, and charismatic, as if cloning were a form of Xeroxing all aspects of a human being's physical, mental, personality, and social traits. This is, of course, a highly misleading analogy. We have the evidence of natural clones – "identical" twins. Not only do identical twins have the same genetic structure, they are usually raised in as close to the same environment as humans can be raised. Yet though they may end up similar in many ways, they also end up dissimilar, in indefinite numbers of other ways. Certainly, they are not the same person, thinking the same thoughts at the same moment.

This is a fortiori true of artificial clones relative to the cell donor from whom he or she is cloned. First, in *all* clones a host of random and prenatal factors will create differences in gene expression that ramify in significant phenotypic differences. Thus, when reproductive physiologist George Seidel cloned the first Holstein calves by splitting a developing embryo, people were surprised to find that the pattern of black and white in the twins was significantly different. Second, and even more important, we know through both science and ordinary experience and common sense that a person is shaped by both heredity and environment. What we are is underdetermined both by our genes alone and by our experiences alone. The child of highly athletic parents raised as a bookworm and couch potato is likely to turn out less athletic than the child of athletically inept parents whose body is trained in a variety of athletic pursuits from infancy through maturity. And most assuredly, an adult who clones him- or herself will inevitably be unable to provide an environment for the clone similar to that in which he or she grew up thirty or forty years earlier, thereby diminishing the likelihood of similarity of the resultant child to the parent.

Let us suppose that I am cloned (something my colleagues have frequently suggested, given my type A personality, my tendency to overcommit, and my cavalier disregard for calendars, which has led me on occasion to promise to be in two places at once). Tired of these embarrassing scenarios, I decide to create a clone of myself and, as tomorrow's vernacular might have it, "do a Dolly." Let us further assume that I have had my genetic material inserted into an egg, and it is developing happily in a surrogate mother's womb. A variety of forces will influence gene expression, including random factors and the prenatal environment. Thus, which genes express, and to what extent, will vary from clone to clone. In this way, variation will be assured from the earliest stages of development.

So from birth and even before, all clones will be different from each other and different from the parent organism. Differentiation will continue to occur as little Bernie grows in virtue of environmental or natural considerations. For example, whereas I was raised in New York, Bernie Jr. will be raised in Colorado. Whereas I was raised with one parent and a grandmother, Bernie Jr. will be raised by two parents, one of whom is me. If my wife hates the idea of a cloned Bernie, that will certainly shape the child's emotional life, as my wife's cooking will shape his physical development. Although my mother discouraged my participation in sports, I will bestow a set of weights on Bernie Jr. while he is in the crib, and so on. In fact, my identical twin – if I had one – is far more likely to be similar to me than my artificial clone, since the twin and I experienced more or less the same environment. Yet that situation did not make us the same person.

Obviously, the same logical point holds of the cloned Hitlers. Hitler was a product of an elaborate constellation of historical events. He was raised in anti-Semitic Austria, and worked his evil on Germany, a country with a long history of seeing the Jew as a pathogen in the body politic. Further, Germany had lost a complex war; anarchists and left-wing revolutionaries ran rife, opposed by disenfranchised, right-wing, ex-soldiers; the Allies imposed draconian terms on the defeated Germany; inflation reached unprecedented heights; Hitler was rebuffed in his attempts to become an artist; his father was a very strict authority figure; and so on. A cloned infant Hitler raised in Wyoming ranch country, Upper West Side Manhattan, or even today's Germany would not be in a position to actualize the abilities that made him a dictator, even assuming that these abilities were strictly genetically determined, which they most likely were not. A young Hitler given art lessons and choice opportunities in art might well blossom in that direction; one whose debating skills were nurtured may well have become a lawyer.

Again, the same logic prevails regarding the other favored media example: cloning a basketball team of Michael Jordans. In the first place, the cloned Jordans might not even be authentically interested in basketball, might excel in some other sport, might pursue no sport, or might well rebel against the reason they were created, even as many children rebel against what their parents push them toward. Further, a team of Jordan clones would, even if greatly interested in basketball, probably not get much better than Jordan. If we wished to significantly progress beyond Jordan's genetic abilities, we would probably do better to have him impregnate some world-class women basketball players, and pay the mothers to raise the children in an environment focused on basketball, though here too there is no guarantee we would create ball players.

In any event, it is difficult to see what dangers would emerge from cloning individuals, even "bad" individuals, given that how they turn out is open to chance, even as how natural children turn out is open to chance. Surely, the majority of people would not clone themselves. So what harm is there in freezing a small portion of the gene pool? Again, even if large numbers of healthy individuals were cloned, what danger would this pose to the gene pool? Since the bulk of human adaptation is technological, rather than biological, it is difficult to see what harm a large number of clones would cause, assuming they were possessed of normal abilities.

In my view, a far greater threat lies in our current approach to genetic disease. At present, largely out of fear of a religiously based mind-set of "there are certain things humans were not meant to do," the strategy for fixing genetic diseases is to attempt to supply the missing gene product – so-called somatic cell therapy – rather than to repair the defect at the genomic level. The result of such an approach – if successful – is to proliferate defective genes in the human population whose deleterious effects are masked by therapy. In the event of nuclear winter or other natural disasters disabling our technological capabilities, vast numbers of people will be unable to survive. I see no comparable danger in a cloned population.

One persistent claim regarding the alleged risks of cloning concerns the negative impact it is likely to have on the nature and structure of the traditional family. In the words of the National Bioethics Advisory Commission:

> In recent years, there has been a heightened awareness of the immense damage done to a child and to a society when fathers are missing from the home. In Washington, D.C., homicide is a leading cause of death for young men; there are many factors that contribute to the pattern of violence, but fatherlessness is among them. Cloning brings children into the world with not one but two missing parents. Genetically, the parents of a cloned child are the parents of the cell donor; they may have been dead for 30 years. Further, the parents are missing or hard to identify not because of specific tragedies but as a matter of planning.
>
> Further, with a cloned child, the basic human relationships need to be redefined. Is the cell donor your brother or father? Is the cloned child a sister or a daughter?[9]

Though piously couched in invoking echoes of "family values," there is little content to this argument. "Missing fathers" is a major pattern of behavior in certain subcultures, where it indeed may be part of the causal conditions for violence, but the same pattern of evidence does not occur in some other subcultures with missing fathers. Furthermore, if a married couple cloned the husband or wife (or someone else) as an answer to childlessness, it is manifest that the children would indeed have two parents. Though cell donors may be deceased, so too may be sperm and egg donors.

Cloning is no more necessarily erosive of family life than are a wide variety of other parenting modalities. Does adoption erode families? How about cross-racial adoption? Does going to a sperm bank or fertilizing a donor egg? If gay or lesbian couples have children using assisted reproduction, is it similarly an attack on the sanctity of the family? Is the child equally confused about who the parents are? Perhaps so, but there has been no comparable bioethical outcry in that area. Indeed, if one did raise the same concerns about same-sex parents, one would be rapidly denounced as homophobic, and we have seen no presidential panel raising these concerns. Instead, one sees only sectarian religious objections.

[9] NBAC, "Summary of Arguments against Cloning," p. 2.

What of single parents who adopt or use assisted reproductive technology to have a child? What of the huge percentage of marriages ending in divorce? What of illegitimate children? These surely pose as great a threat, or probably a greater one, to the traditional family, as does cloning. But where is the outcry to legislate an end to divorce?

Parenting is a functional role. Certainly, parenting is most often associated with permanent, identifiable, biological mates, but it need not be so, as all of the above examples well illustrate. Adoptive parents, single parents, or gay parents may function better as parents than "normal" married couples. Even animals will "parent," or "adopt," other animals, sometimes animals of a different species. So what!? People with cloned offspring may treat them well or badly, be good or bad parents, but that will not be decided by whether or not the child is cloned.

One can similarly dismiss the NBAC "argument" that cloning is an assault on the "dignity of human procreation." As in so many "human dignity" locutions, I have a hard time understanding what that means. As many literary figures have pointed out, there is little "dignity" in the sex act, whether performed by humans or ostriches. If the sex act is so "dignified," why do so few official portraits or statues depict the relevant "dignitary" *in flagrante delicto*? If there is "dignity" there at all, it lies in how one shoulders the burden of parenting, not in a particular biological route to having a child. Is there more "dignity" in poor Third World slum dwellers who have a child because they don't understand birth control than in an educated person who chooses to implant an embryo derived from cells of a deceased infant? (I return to this case shortly.)

The same sort of objection can also be raised against the NBAC report's claim that cloning represents "an assault on the dignity of the conjugal union." The report says:

> The existence of children is a persistent sign of the parents' mutual love, and an invitation to ponder the endless mystery of gender differences. The potent and universal sign of hope is blurred or lost when the child's life begins in a laboratory.[10]

[10] Ibid., p. 1.

Surely, the first claim is false. Children may be a "sign of mutual love," but they can also be a sign of lust, drunkenness, ignorance of conception, apathy, rape, immaturity, and so on. And the claim that "hope is lost when life begins in a laboratory" is a classic example of the genetic fallacy. A child is still a child regardless of how it is occasioned, whether in loving bliss, teenage revelry, or in a laboratory.

It is particularly ironic to me that the tempest about human cloning was roughly contemporaneous with the announcement that a U.S. woman who had been taking fertility drugs gave birth to septuplets. She was not confronted with social outrage; instead, she became a cultural hero and was showered with positive publicity, money, and gifts. Yet it is far easier to envision a negative or tragic outcome to her story than to a couple who quietly cloned a child. Without external interference, the latter can set about the task of parenting, while the other woman (or couple) are playing against a stacked deck. As any parent of triplets will tell you, parenting three infants simultaneously presses human beings to their limits – how, then, can one manage seven? I would venture a guess that a cloned child not labeled a clone has a better chance of a good childhood than a child who is one of a set of septuplets. In any case, one cannot decide this sort of issue a priori, which is precisely what those who reject cloning would do.

Another risk-related argument evolves around the claim that cloning "smacks of eugenics," a claim that is little more than guilt by association. Granted that eugenics has in fact been associated with much evil at the hands of the Nazis and even in the United States: forced sterilization, euthanasia of the "unfit," and so on. That does not mean that all genetic improvement is conceptually connected with evil. Who would object if we could, through genetic engineering with no untoward consequences, ablate the gene or genes for cystic fibrosis, Lesch-Nyhan's syndrome, Cruzon's syndrome, and other hideous genetic diseases? That is surely "eugenics," yet seems to be a paradigmatically good sort of thing to do, leading to a better universe. To stop genetic improvement of this sort because Hitler or others did evil things under the same rubric is to succumb to the most simplistic sort of slippery slope argument – "today we ablate cystic fibrosis; tomorrow we make everyone genetically Caucasian." No – we ablate cystic fibrosis and guard against making everyone Caucasian.

This brings up a related issue. Some minority group members fear cloning as inherently connected with racism. For example, African American medical ethicist Marian Gray Secundy affirms:

> Ethnic Americans are extraordinarily suspicious of *any new* scientific technologies. This is particularly true for, but not confined to, the African-American community.... The prevailing sentiment is that scientists cannot be trusted, that white scientists particularly are dangerous, that abuses are inevitable and that all manner of evil can and will most likely be visited upon the most vulnerable, e.g. ethnic groups and the poor. A family practice resident commented, "The White Man has a God complex." Others raised concerns about possible abuses, among them that "they" clone soldiers for war, making "us" subservient tools.[11]

On the other hand, other representatives of ethnic groups embrace cloning. In the same publication, the Reverend Abraham Akaka, a Hawaiian Native American minister argues:

> For aboriginal people of our planet who see themselves as dwindling and endangered species, cloning of the best of their race will be a blessing – a viable avenue for preserving and perpetuating their unique identities and individualities upon lands they revere as father and mother; a way to extend their longevity on earth.[12]

A final argument regarding danger inherent in cloning is extremely widespread, yet almost impossible to take seriously. This is the notion that people will clone themselves in order to create spare parts – hearts, lungs, livers, and so on. Despite our encountering this concept in medical thrillers, this is truly a stupid idea. For one thing, one would need to wait until the clone reached adulthood – almost two decades. For another, as I discuss below, animals whose organs have been genetically engineered to be compatible with the human immune system – xenotransplantation – are close to realization. (This, of course, raises a separate moral issue of animal exploitation.) Third, such use of a human is clearly covered and forbidden by our current social ethic – it would be premeditated murder. Society would not allow this any more than one is allowed to sell one's children for their organs. Fourth, it would be enormously inefficient to raise people for spare parts.

[11] Woolfrey and Campbell, *Reflections*, p. 5.
[12] Ibid., pp. 2–3.

What seems much more likely to me is the development of a technology whereby one can clone "spare organs" from the relevant tissue, rather than growing and discarding the entire person. Such organ cloning seems in principle possible, and research in this area is in fact under way.

Having examined the alleged risks of cloning and found them to be trivial or at least no greater than those of other technologies – including reproductive technologies – that we already accept, it remains to consider possible benefits of cloning. Let us begin by citing a case that strikes most people with whom one discusses it as plausible: Imagine a couple who have struggled to have a child naturally. The child is born, and they can never have another. The child, at one month of age, is struck by lightning in a freak accident and is dying and irreversibly comatose. The child has not developed a personality yet, and they wish to replace him. Would they be justified in cloning another child from the dying child's tissue? (It's not as if the new child would always be measured against the idealized and deified dead child – that would create moral problems in terms of unfair demand on the new child.) Basically, they want to have a child of their own. They are not cloning themselves, only the dying baby. Intuitively, few of us would say they are *morally wrong* to create the clone, though we might still fell some squeamishness.

Second, we have already encountered the sort of case raised by Rev. Akaka, wherein cloning would be used to save a people whose numbers have been decimated in one way or another. Though the suggestion is staggering, I see no moral issue in cloning victims of genocide, assuming the children would be placed in proper families and so on. The cloning of Jewish Holocaust victims whose genetic material we had, for example, while certainly eerie, would not be morally problematic, but rather a reasonable attempt to replenish the Jewish gene pool so cruelly robbed. (Here, of course, we assume that we can clone effectively from deceased people or perhaps that the Nazis had saved cell lines.)

Third, as is in fact discussed in the NBAC report, we can imagine parents whose infant requires, say, a bone marrow transplant, cloning that infant *in part* to provide the marrow, assuming, of course, that the parents genuinely want another child. Such situations have already arisen with parents in this sort of situation electing to have another child in part to save the first, ailing one.[13] It is difficult of course to judge parental motivation in having a child, but that is the case for normal reproduction as well.

Fourth, a celibate person may wish to pass on his or her genes. It is hard to see why this is morally unacceptable if gay couples or single people may morally use assisted reproductive technology to reproduce.

Fifth, one can imagine a devastating disease that wipes out large portions of the population but to which a small percentage of people are naturally resistant. Surely, it would be reasonable and acceptable to clone such people so as to assure the further spread of that resistance in the population, assuming that the people were willing to be cloned.

Thus, it seems to me that there are possible scenarios where cloning might be beneficial to individuals and to society, and free of risk, or at least where it would be a plausible reproductive strategy with only a religious or very weak "intrinsic wrongness" argument against it.

Our third category of issues now arises, namely, whether cloning is likely to have a negative effect on the well-being of cloned beings, animals, and/or people.

The first such possible consequence arises out of the possibility that cloning per se can have unexpected and deleterious effects on cloned people or animals. Although one is putatively creating an organism that ought to end up indistinguishable from a naturally derived animal, it is conceivable that the process of cloning could itself have deleterious effects that emerge at some stage in the life of the organism. This phenomenon has already been manifested in cattle clones created by splitting embryos by nuclear transfer. According to veterinarians working with these animals, they have been oversized and thus difficult to birth, had difficulty surviving, and have also been behaviorally retarded, requiring a good deal more care at birth than normal calves.[14] (Indeed, there seem to be many problems in noncloned animals or people created by in-vitro fertilization.) The cause of this is not known, and it is quite possible that clones such as Dolly could "crash" later in life in virtue of some unknown mechanism. At this point, there

[13] NBAC, "Summary of Arguments against Clonings," p. 55.

[14] Garry et al., "Post-Natal Characteristics of Calves Produced by Nuclear Transfer Cloning."

is no evidence for this concern – it is an empirically testable possibility that will be verified or falsified as our experience with cloned animals develops. If it turns out that there are in fact unanticipated welfare problems for animals that are cloned, this should and likely will abort the technology until the problems are solved. There is a precedent here from the unanticipated consequences of inserting the human growth hormone gene into pigs, as discussed above. While these transgenic animals were leaner and faster growing as expected, as discussed earlier, they also displayed totally unexpected and disastrous effects, including lameness, synovitis, cardiac disease, ulcers, reproductive problems, and kidney and liver problems.[15] In sum, cloning should not create pain and suffering in people or in animals, and our emerging social ethic for animals clearly has the minimizing of animal suffering as its major thrust.[16]

There is a more subtle sense in which cloning can conceivably create problems for animal welfare and thus give rise to genuine social moral concerns. It could be argued that cloning feeds directly into a view of animals that has led Western societies to demand a higher moral status for animals. I am referring to what my colleague, Michael Losonsky, calls the *commodification* of animals or of life. As we discussed earlier, for most of human history the major use of animals by humans was agriculture – food, fiber, locomotion, and power – and the essence of agriculture was *husbandry*: care and stewardship. Etymologically derived from the Old Norse *hus/bond* (=bonded to the house), husbandry meant putting animals into an environment for which they were optimally suited by nature and augmenting their natural ability to survive and thrive with additional food, shelter, medical attention, protection from predation, and so on. Animal behaviorist Temple Grandin calls this the "ancient contract," wherein both humans and animals were better off in virtue of that relationship. In husbandry-based agriculture, no rational human would hurt his or her animals in any prolonged or extended way, for he or she would destroy the animal's productivity and ultimately harm him- or herself. Husbandry agriculture was thus about putting square pegs in square holes, round pegs in round holes, and creating as little friction as possible while doing so.

[15] Pursel et al., "Genetic Engineering of Livestock."
[16] Rollin, *The Frankenstein Syndrome.*

In the mid-twentieth century, however, humans broke the ancient contract with animals through emerging technology. We could now put animals into environments for which they were not biologically suited, yet keep them productive through technological "fixes" such as antibiotics, vaccines, and hormones. Although the animals surely suffered in such environments, their productivity (or more accurately, the productivity of the operation) was essentially unimpaired. Instead of worrying about the whole animal, we could ignore needs unrelated to productivity. With technological sanders, we could fit square pegs into round holes.

Animal husbandry became animal science, defined as the application of industrial methods to the production of animals. In my view, this is the main factor that has called forth ever-increasing and international social concern about animal treatment. But whether I am correct or not, this modern approach inexorably leads to viewing animals as production machines, rather than as living things with lives that matter to them. Instead of husbandrymen, industrialized agriculture utilizes businessmen and managers, people for whom the financial reward is a significantly higher value than the way of life so central to traditional agriculturalists.

In such an agriculture, animals are products or commodities: pounds of pork, or eggs per cage. The ability to clone them, one might argue, augments and reinforces this view. After all, cloned animals are manufactured and, like cars or soup cans coming from assembly line, seem to be "identical."

I have some sympathy with this concern, the same concern that informs animal advocates' vigorous opposition to patenting animals. But the issue here is far more basic than cloning – it is the industrialization of animal agriculture and the loss of the ethic of husbandry. On traditional hog farms, for example, sows had *names* and received individual attention. In today's huge production units, they do not. Cloning per se is perhaps a reflection of this industrialization, but there is no necessary connection between the two. After all, one can imagine a strongly husbandry-based agriculturalist caring a great deal for his herd of cloned pigs. Just because cloning has emerged from a questionable mind-set does not mean that it could not thrive in a highly morally acceptable agriculture. Just because cloning is a spin-off from industrialized agriculture does not mean that it is conceptually incongruous with sustainable husbandry. Western ranchers – the last husbandry agriculturalists – will continue to provide husbandry for their animals whether they are produced by artificial insemination, cloning, or natural breeding; after all, they are still animals under our care.

It could perhaps be claimed that cloning will accelerate public apathy about farm animal treatment, based on the psychological fact that the more identical beings that exist, the less we care about them, a fact the Nazis exploited in camps.

I can imagine people wishing to clone their beloved dog or cat or horse – companion animals – even as some people preserve their pets' bodies by taxidermy or freeze-drying. I see such a use of cloning as unproblematic and self-limiting. It is not problematic because we will never have monoculture of pets – there are too many different preferences. Cloning would certainly not cause as many welfare problems as breed standards do today, wherein we perpetuate hundreds of genetic diseases in dogs by our aesthetic predilections. Indeed, I see no welfare problems here unless cloning has the adverse effects we mentioned earlier. Such cloning, I believe, would be self-limiting, when people realized that the clone is not their beloved Fifi and may be very different. At that point, people will realize that they may as well just get another puppy from Fifi's bloodline, if she is purebred.

A trivial problem associated with cloning companion animals is the risk of con artists claiming to be able to reproduce one's beloved pet. This has in fact already occurred. Though the scientists in question explicitly state that they are not making a copy of the pet, they play to this desire. Caveat emptor!

The last moral issue we must address regarding cloning of humans concerns possible harm to the clone – what we have called in our Frankenstein typology "the plight of the creature." If, for example, clones inevitably develop some wasting disease with the advent of puberty, cloning would clearly become a morally unacceptable reproductive modality. This is the most powerful reason against anyone's undertaking to clone a human at this stage of knowledge and explains why even strong advocates of cloning research were furious at Richard Seed's irresponsible announcement that he would begin cloning shortly – it is the same concern that greeted the vastly premature attempt to provide a baboon heart to Baby Fae.

Is cloning likely to engender untoward effects in a cloned person? No one knows, as we indicated. As we saw, cloned calves created by splitting embryos were oversized, required birthing by caesarian section, and were behaviorally retarded. By the same token, there are suggestions that any assisted reproduction techniques involve more risks for the resulting organism than normal reproduction does, ranging from problems with infections to possible intellectual impairment. As te Velde and coworkers remark in a commentary in the *Lancet*:

> Manipulation of human genetics or embryos and intervening with the natural process of conception may induce subtle, complex, and far-reaching changes in the genetic material of the offspring and perhaps also of the next generation. Before new assisted reproduction techniques are adopted as routine treatment for infertility, they should be assessed extensively in animal and human embryo research, then in clinical trials, during which the children must be monitored long term. Lessons learned from the unexpected effects on fetal development of drugs that were not adequately assessed before introduction should apply here. Scientists and governments should be discussing how the introduction of assisted reproduction techniques should be regulated.[17]

Obviously, our discussion must be focused on a scenario where sometime in the future, cloning has been perfected in animals, and the technology proceeds as smoothly as any other assisted reproductive technology such as surrogate motherhood or in-vitro fertilization. To attempt cloning today, via the technology that produced Dolly, would clearly violate established medical-ethical principles, since hundreds of attempts would be required, and risk of late-term spontaneous abortion threatening the life of the mother would be significant. Furthermore, we are not yet in a position to judge the potential deleterious consequences to the cloned child, and thus, again, it would be immoral to proceed with cloning in a cavalier fashion. The requisite certitude can be achieved only after years of research. (We are also here holding in abeyance the question of whether the requisite animal experimentation is morally acceptable, as it almost certainly will proceed.)

Assuming we reach a point where the process of cloning does not damage the person so created, are there any other risks to the clone to worry about? There is certainly a risk of societal revulsion at clones.

[17] Te Velde et al., "Concerns about Assisted Reproduction."

On the other hand, such revulsion will probably diminish after it is revealed that some beloved figure, say, a movie star or Mother Theresa, was in fact generated by way of cloning. Finding out that someone was cloned well after the fact, when we already love and admire them, can well defray our revulsion. On the other hand, labeling a child as a "clone" from birth and marginalizing him or her will almost certainly serve as a self-fulfilling prophecy, making that person "not quite like us" or "freakish." Obviously, such a label would warp a nonclonal child if he or she were so identified, as well as one derived by cloning.

Thus, one genuine risk to a cloned human grows out of his or her possibly being identified and labeled by others as monstrous, or less than human. This, of course, is not a risk of cloning per se, but of social reaction to it. If religious leaders led a witch hunt to "ferret out soulless clones," clearly, cloned humans would be at risk, though of course it would equally be a risk to anyone misidentified as a clone. The best way to guard against this risk is to maintain strict confidentiality about who is cloned until such time as cloning is socially accepted or at least widespread.

Given environmental influences, pre- and postnatal, age differences, difference in hair and clothing styles, and so on, no one could recognize my clone as a clone, rather than as my normally derived progeny or grandson, were I older when he was cloned. Assuming no untoward effect of cloning, neither he nor anyone else would ever need to know that he was clonally derived, any more than children need to know they were adopted. To say that cloning is unacceptable because cloned children will be unfairly stigmatized for no good reason begs the question against cloning – the plausible response is to assure that they are not so marginalized.

Chicago physicist Richard Seed's headline-grabbing announcement that he would clone a human being fueled the political flames growing out of the bad ethics hysteria surrounding cloning. Between 1998 and 2004, no fewer than thirty-four human cloning–related bills were introduced in the House of Representatives to prevent such cloning, leading to a total number of forty-four bills directed at this issue. Fourteen European countries ratified a Council of Europe prohibition on reproductive and therapeutic cloning. (I discuss therapeutic cloning below in the section on stem cells.) The flames were further fanned in December 2003 when the religious cult known as Raelians announced the cloning of an American woman but provided no evidence for the truth of this claim.

One novel use of biotechnology stems from the significant need for organs for transplantation. According to the Health Resources and Services Administration, the federal agency overseeing transplanting, as of 2001, some 75,000 patients need organ transplants, a number that is increasing, yet donors numbered roughly 10 percent of the per annum.[18] One suggestion that appeared promising was to use genetically modified animal organs, notably hearts and kidneys, from pigs that were rendered immunologically compatible with the human immune system to avoid rejection. The pig is roughly the right size and body shape and is relatively easy to raise and breed; by the mid-1990s many research groups were involved in studying the possibility of xenotransplantation.

It is relatively easy to utilize the categories we have developed for analyzing genetic engineering to examine xenotransplantation. One need not be Nostradamus to anticipate reaction of "there were certain things humans were not meant to do" when faced with such a technology. The idea of a pig part put into a human being, particularly a heart, which many people believe to be the physical counterpart of the soul, was a priori unacceptable to some, particularly the religious. In a similar vein, xenotransplantation evokes the "playing God" objection. Neither of these objections is rationally defensible, as "pigness" is not transferred with pig parts. If it were, all those who had received pig heart valves over the past fifty years would be marked as porcine. And "playing God" would surely apply to much of human technology.

The second category of concern – dangers inherent in the technology – is real and serious. Inherent in the technology is the possibility of transferring harmful viruses from pigs to humans or, more unpredictably, of pig viruses changing when introduced into human hosts and becoming pathogenic and perhaps lethal. This specter of "xenozoonoses" caused by unpredictable recombination of viruses in the person receiving the transplant can also conceivably lead to epidemics of untreatable viral disease. For these prudential reasons, the Council of Europe imposed a moratorium on xenotransplantation in

[18] Secretary's Donation Initiative, http://www.organdonor.gov/secinitiative.htm.

1999, and enthusiasm for this technology has significantly diminished in the United States. This is a serious concern, and one cannot envision a future for xenotransplantation until it is resolved.

What of the "plight of the creature," that is, the issues that arise in terms of the well-being of the animals engineered for organ transplantation, particularly in terms of what I have termed the "new social ethic" for animal treatment emerging in Western societies over the past three decades? When it comes to animals used for food or for biomedical research and safety testing, the ethic overwhelmingly concerns itself with how the animals live and how they are cared for, not with animal death (unless it is painful).

A striking anecdote illustrates this point well. During the 1980s, a scientist colleague of mine at a British university was convinced that the laboratory animal laws passed in the United States and Britain in 1985 and 1986, respectively, were each by themselves incomplete. While the British law was a system of licenses and inspections, the U.S. law was based, as discussed, in animal care and use committees and protocol review. Enjoying considerable credibility at his university, my friend persuaded his peers to voluntarily adopt the U.S. committee system as an addendum to what British law required. A few years later, he had another idea, arguing that some animal experiments were not worth doing even when they merely involved euthanasia of the animals. He proposed a calculus designed to balance painless animal death against scientific value. His colleagues reacted so negatively that he in essence felt compelled to move abroad.

Clearly, pigs or other animals raised for xenotransplantation will be well treated if only to minimize stress, and the organs harvested under strict aseptic conditions with terminal surgical procedures involving state-of-the-art anesthesia. A pig used for xenotransplantation will almost certainly live a better life than a pig used for food raised under confinement conditions, and its death will certainly be a far better one. It is difficult to believe that a society that accepts the killing of pigs for bacon will cavil at their more painless killing to save human lives. The most that may occur would be social demand that the transplantation pigs live decent lives under conditions congenial to their natures, which much of society has in fact already demanded for pigs used for food. Given the high economic value of these pigs compared with pigs used for food, the expense of creating exemplary living conditions for them does not pose a major issue. Thus, animal welfare issues do not loom large in this discussion.

One final area of biotechnological exploration that has also occasioned a storm of "ethical" debate is embryonic stem cell research in human beings. Embryonic stem cells are cells that appear in early embryonic development, at the stage known as the *blastocyst* (three- to five-day-old embryos). These cells are at this point undifferentiated, but soon develop into the specialized cell types that make up the tissues and organs of the body: heart, lungs, nervous system, and so on. Theoretically, these stem cells can be removed from the embryo and transplanted into people suffering from various degenerative diseases who might be helped by new tissue growth. The scientific challenge is to control appropriate differentiation into tissue. Probably the earliest promise has been shown of embryonic stem cells differentiated into dopamine-producing neurons to treat Parkinson's disease.

Embryonic stem cells were removed from embryos in 1998 by James Thomson at the University of Wisconsin. Thomson proceeded to establish five cell lines, wherein populations of embryonic stem cells continue to grow and divide in culture and remain undifferentiated. Thomson's stem cells were taken from surplus embryos that were derived from in vitro fertilization.

It has been discovered that there also exist adult stem cells, whose function is to differentiate into the various cell types in the organs and tissues where they reside and repair damage. Adult stem cells are found in neural tissue, bone marrow, and elsewhere. Unlike embryonic stem cells, adult stem cells generally cannot be reproduced indefinitely. There is, however, some evidence that adult stem cells retain a certain degree of plasticity, in that they can develop into various types of tissue. Bone marrow stem cells, for example, can produce skeletal muscle, cardiac muscle, neural tissue, and liver cells.

Embryonic stem cells are preferred for medical research both because they can differentiate into all cell types and because they are easily grown in culture. In addition to Parkinson's, they can theoretically be used to treat tissue damage resulting from Alzheimer's disease, stroke, burns, heart disease, arthritis, and diabetes; anywhere that defective or injured cells must be replaced.

To many scientists and patients, stem cell research seems like a scientific Holy Grail, offering cure potential for hitherto untreatable diseases. Furthermore, we can proliferate these embryonic stem cells indefinitely and do not need to harm humans or animals to acquire them. Yet this technology has generated a debate as hot as those attendant to any new technology that has even divided politically conservative Republicans. President George W. Bush banned federal money being spent on stem cell research, with the exception of work on extant human embryonic cell lines. Senator Orrin Hatch vehemently supports such expenditures, as do patient advocacy groups. Britain has moved to encourage stem cell research and is attracting some of the best U.S. scientists in the area. And all of this furor is over an "ethical issue" that in my view falls into the category of spurious ethical issues, the category of "certain things humans were not meant to do" or "cloning violates God's will."

For a certain subgroup of society, simplistically referred to as the "religious right," the major ethical issue in embryonic stem cell research grows out of the fact that to harvest these cells, one must destroy the embryo. This has gained significant opposition from pro-life, anti-abortion groups, including the Catholic Church with Pope John Paul II himself strongly opposing the creating of embryos for harvesting cells. As the Pope put it in a speech given on July 23, 2001:

> Another area in which political and moral choices have the gravest consequences for the future of civilization concerns the most fundamental of human rights, the right to life itself. Experience is already showing how a tragic coarsening of consciences accompanies the assault on innocent human life in the womb, leading to accommodation and acquiescence in the face of other related evils such as euthanasia, infanticide and, most recently, proposals for the creation for research purposes of human embryos, destined to destruction in the process.
>
> A free and virtuous society, which America aspires to be, must reject practices that devalue and violate human life at any stage from conception until natural death. In defending the right to life, in law and through a vibrant culture of life, America can show the world the path to a truly humane future in which man remains the master, not the product, of his technology.[19]

[19] Pope John Paul, speech delivered July 23, 2001.

In other words, on this view, human life begins at conception, a clear replication of the anti-abortion argument. A very well-reasoned paper by Ray Bohlin of Probe Ministers, a fundamentalist group, affirms that

> these tiny embryos are already of infinite value to God. We're not going to redeem them by killing them for research. Each embryo is a unique human being. . . .[20]

In essence, this is a Protestant version of the Pope's claim.

It is a perfectly respectable theological position to affirm that human life begins at conception, presumably with the infusion of a soul into the embryo. But, at least in theory, theological positions do not serve as a basis for secular social ethics in Western democratic societies; hence the constitutional provisions separating church and state. That is not to suggest that religiously based ethics cannot come to dominate social ethics – many of the opponents of capital punishment draw their ides from religion – but rather that something should not become socially forbidden simply because it violates certain theological precepts; it must have some rational ethical base to justify a ban.

The question of when an embryo becomes a full human being, like the question of when death occurs, is not a factual question to be decided by science but a social decision. Science is *relevant to* that decision, for example, when it informs us that unborn children are likely to feel pain by the third trimester, but it does not dictate that decision. In our current social ethic, the decision has been made that the fertilized egg is not a human person. This decision is manifested in our social ethic on abortion, which essentially treats the embryo as the mother's property, and she can, until birth, dispose of it as she sees fit. Similarly, in vitro fertilization facilities create more embryos than they need and dispose of them freely and legally. Thus the view that human life beings at conception is not adopted by our social ethic.

To be sure, we can rationally criticize our social ethic. For example, in my own personal view, given what we know of embryonic

[20] Bohlin, "Controversy over Stem Cell Research."

development, late-term abortions are morally questionable, being closer to killing a baby than disposing of an unwanted part of a mother's body. Scientific information is highly relevant to pressing forward such a position.

But, as yet, no one has given a strong secular argument to persuade us that a blastocyst is a fully formed human and worthy of being treated as such. We have no reason to believe that such an early stage embryo can think, feel, or exhibit features constitutive of a sentient animal, let alone a human. What is *potentially* a human being, in society's consensus view, is not yet a fully human being who bears rights. Though this violates some versions of Christianity, it is in complete accord with other versions and with Judaism, where humanity occurs at birth.

The debate on whether human life begins at conception is inherently not one to be resolved by science or by ethics; rather, it is a metaphysical issue, an issue of how we look at and categorize the world. Both pro-abortion and anti-abortion proponents would be likely to accept the ethical norm "One should not kill human beings for convenience." The operative question is the metaphysical one, "What is a human being?"

Thus, given the fact that a mother may dispose of an aborted fetus as she sees fit, clearly the logic of our social ethic does not accord with the Catholic view. The logic of the social ethic would be, with appropriate informed consent, to allow the aborted fetus or discarded, in vitro–fertilization embryo to be used as a source of stem cells. Proponents of the view that humanity begins at conception are perfectly within their rights to advocate a new, religiously based metaphysical position but seem to overstep rational ethics and move more into politics when they use political clout to block stem cell research. That is why I consider this debate a spurious ethical issue, though very much a live political one.

The issue has been exacerbated by work done from 2001 to 2004 when scientists developed a technique for cloning human embryos, which could be used as a source of embryonic stem cells. This technique, known as "therapeutic cloning," has further polarized the pro- and anti-embryonic stem cell camps. While almost everyone supports a ban on "reproductive cloning," that is, creating full-blown humans by cloning as Dolly was created, those who support embryonic stem cell research support "therapeutic cloning," while those who believe life begins at conception do not.

Our lengthy discussion of biotechnology and ethics is important. Unless science engages the ethics of biotechnology, we as a society run a great risk of losing a powerful tool that can enrich our lives immeasurably to public rejection driven by bad ethics.

9

Pain and Ethics

Thus far, we have examined the relationship between scientific ideology and the neglect of major ethical dimensions of science, largely caused by the component of scientific ideology that declares science to be "value free" and "ethics free." But while the explicit denial of values is certainly going to be the most obvious cause of ethical neglect, we cannot underestimate the more subtly corruptive influence of the second component of scientific ideology we have delineated, the denial of the reality or knowability of subjective experiences in people and animals.

Obviously, concern about how a person or animal feels – painful, fearful, threatened, stressed – looms large in the context of ethical deliberation. If such feelings and experiences are treated as scientifically unreal, or at least as scientifically unknowable, that will serve to eliminate what we may term a major call to ethical deliberation and ethical thought. Insofar as modern science tends to bracket subjectivity as outside its purview, the tendency to ignore ethics is potentiated. For example, in our discussion of animal research we have alluded to the absence of pain control in animal research until it was mandated by federal legislation.

While this is certainly a function of science's failure to recognize ethical questions in science, society in general, except for issues of overt cruelty, also historically neglected ethical questions about animals. Ordinary people, however, were comfortable in attributing felt pain to animals (a matter of ordinary common sense) and adjusted their behavior accordingly, even in the absence of an explicit ethic for animals being prevalent in society. Scientists, however, in being trained to disregard ordinary common sense regarding animal subjective experience, were not moved by what nonscientists saw as plainly a matter of pain. Hence one of my veterinarian colleague's response to my concern about howling and whining in his experimental surgery dogs: "Oh, that's not pain, it is after-effects of anesthesia."

In other words, the denial of the reality (or at least scientific knowability) of pain in animals provided yet another vector for ignoring ethics, since ethical concern is so closely linked to recognizing mental states. I shall, surprisingly, document below a similar problem in human medicine.

It is certainly the case that modern science (i.e., the science beginning with Galileo, Descartes, and Newton) began with preconceptions uncongenial to taking subjective experience as part of scientific reality. Medieval and Aristotelian science (medieval science was at root Aristotelian) set itself the task of explaining the world of sense experience and common-sense experience, a world of qualitative differences to which sense experience provided largely accurate access that, when it failed, could be corrected by additional sense experience. As I have explained elsewhere,[1] Aristotelian science took the position that the world of ordinary experience was the "real world," that "what you see is what you get." Indeed, that in my view was one reason that the Aristotelian world-view lasted so long – it was based in and congenial to ordinary experience.

This is, of course, not the case with the new science. Everyone who has taken introductory philosophy recalls reading Descartes' sustained attack on the senses and on common sense, which was intended to undercut the old world-view and prepare people to accept that reality is not as it appears to be but as reason and mathematical physics tell us it is. Descartes' program was completed by Newton but retained the same logic.

Thus, as we remarked at the beginning of this book, physical science became the paradigm case for all science, with "objectivity" the primary mantra in all fields. Even the social sciences strived to be "objective." Subjective experiences were strongly disvalued (even though science

[1] See, e.g., B. Rollin, *The Unheeded Cry*.

was said to be based in "experience"). By the 1920s, as I have recounted in detail elsewhere, subjective experiences had been relegated to nonpersons all across science, with J. B. Watson and behaviorism finally eliminating it even from psychology when Watson skillfully sold the idea that psychology, in order to achieve parity with the "real" and successful sciences, needed to become the science of rat behavior and learning. Indeed, Watson came perilously close to affirming that "we don't have thoughts, we only think we do."

As long as I am discussing the scientific revolution, it is important to discuss, as I briefly mentioned in the first chapter, that the scientific revolution itself presents us with a superb example of a change in values. Consider the Aristotelian/common-sense world-view. Can one imagine a crucial experiment that would falsify the claim that the world is best understood in terms of adherence to sense experience, teleological explanations, and qualitative differences, and prove that such explanations must be abandoned in favor of mechanistic, mathematical, quantitative explanations that ignore qualitative distinctions? Since experiments set up in the Aristotelian paradigm would necessarily be qualitative, and since the paradigm determines what counts as relevant data, how could such an experiment ever disprove the paradigm itself? (The same, of course, holds true for overturning an extant mechanistic paradigm in favor of a qualitative one.) Thus the Aristotelian/common-sense world-view was not falsified or disproved; it was rather set aside by virtue of the rise of new values that clashed with it, for example, the belief that God is a mathematician and that under qualitative diversity there must exist quantitative uniformity. The Aristotelian approach was more *disapproved* than disproved.

As mechanism became the regnant conceptual paradigm for physics, its dominance was gradually replicated in other sciences, such as chemistry, biology, and geology. This ideal was enunciated in positivism, which affirmed in the twentieth century that psychology would be reduced to neurophysiology, neurophysiology to biochemistry, chemistry to physics. For those less radical, the move was nonetheless to eliminate the subjective from science, as Watson did with turning psychology to the study of overt behavior. It is again noteworthy that this transformation was not necessitated by experimental evidence or logical analysis overturning the coherence of looking at subjective states. Indeed, as I have shown elsewhere, the alleged historical inevitability of behaviorism as reconstructed in histories of psychology will not stand up to rational scrutiny.[2] None of the major figures in psychology prior to behaviorism disavowed consciousness. In fact, Watson "sold" behaviorism through rhetoric, arguing that only by turning to the examination of overt behavior could psychology become analogous to physics and lead to the ability to manipulate behavior socially, to eliminate criminality and other socially deviant behavior, which, in the end, is learned behavior.

The same pattern of what we may call "physicalization" – elimination of the subjective as irrelevant to science – took place in medicine. Particularly with the rise of molecular biology and sophisticated biochemistry, disease was increasingly seen as defects in the machine, and subjective states as, to use Ryle's apt phrase, "ghosts in the machine." Even psychiatry, by the end of the twentieth century, had come to see mental illness not as "mental" or "behavioral" but as biochemical: insufficiencies or excesses of certain chemicals. The management of such diseases became a matter of balancing an individual's chemistry, not of analysis or individual or group therapy.

Traditionally, before the physicalistic turn, medicine was, and aspired to be, a combination of science and art. The science component came, of course, from its attempt to develop generalizable, lawlike knowledge that would remain invariant across space and time. Such knowledge was sought regarding the working of the body, the nature of disease, and valid therapeutic regimens, though medicine often fell short of the mark in all of these areas. The element of art was patent in medicine. Art deals with the individual, the unique, with the domain of proper names; with the person, not merely the body; with that which does not lend itself to generalization; with the subjective psychological aspects of a person well as with the observable. A physician was thus expected to be both lawlike and intuitive, the latter not in any mystical way, but rather in a manner that is focused on this particular individual and his or her subjectivity and felt experience. And in understanding the individual – who is by definition unique – all information, be it first-person reports or objective measurements, was relevant.

[2] Ibid.

In some ways, the physicalization of medicine was a boon to sick people – there now existed science and evidence-based ways to develop and test drugs and other therapies for safety and efficacy. But, in other ways, it was a detriment. In the first place, how the patient felt became significantly subordinated to how they objectively "were." Medical success came to be measured in terms of how long the patient lived; days alive was an objective parameter that could be quantified, as opposed to the subjective measurement of quality of life.

Cancer medicine provides an excellent example of this view. Oncology was directed at eliminating the tumor and buying a measurable increment in life span, or time before death. Quality of life, suffering attendant on chemotherapy or radiation, loss of dignity in the course of treatment, and psychological and economic toll on family were not measures that scientific medicine was wont to adopt. "Buying extra time" was the goal. And yet, as numerous authorities have told us, patient concern is primarily about suffering, not about death per se.

In general, people who seek voluntary euthanasia do so because they fear pain, loss of dignity (e.g., of the sort that comes from incontinence), helplessness, dependence, and stress on the family.[3] Obviously, they fear such experiences more than they fear death. Yet scientific medicine does not worry about such "hindrances" to prolonging life. In particular, and crucial to our argument in this chapter, felt pain becomes not fully medically real, since it is not observable or objective, or mechanistically definable. In that regard, I vividly recall what one nursing dean told me: "The difference between nurses and doctors is that we worry about *care,* they worry about *cure.*" In turn, recall that the institution that has most concerned itself and done the most for the terminally ill is hospice, and hospice was founded and is dominated by nurses, not physicians.

In 1973, psychiatrists R. M. Marks and E. S. Sacher published a seminal article on pain control in which they demonstrated that almost three out of four cancer patients studied in two major New York hospitals unnecessarily suffered moderate to severe pain because of undermedication with readily available narcotic analgesics.[4] The authors were psychiatrists initially brought in to consult on patients

[3] Callahan, "Death and the Research Imperative."

[4] Marks and Sacher, "Undertreatment of Medical Inpatients with Narcotic Analgesics."

putatively having a marked emotional reaction to their disease. On examination, they determined that the problem was undertreatment of pain that led to the emotional responses, rather than a psychiatric problem. Though their article received a great deal of attention, this disgraceful state of affairs was confirmed by other studies and by an extraordinary editorial in *Pain*, fourteen years later, by John Liebeskind and Ron Melzack, two of the world's most eminent pain researchers:

> We are appalled by the needless pain that plagues the people of the world – in rich and poor nations alike. By any reasonable code, freedom from pain should be a basic human right, limited only by our knowledge to achieve it.
>
> Cancer pain can be virtually abolished in 80–90% of patients by the intelligent use of drugs, yet millions of people suffer daily from cancer pain without receiving adequate treatment. We have the techniques to alleviate many acute and chronic pain conditions, including severe burn pain, labor pain, and postsurgical pain, as well as pains of myofascial and neuropathic origin; but worldwide, these pains are often mismanaged or ignored.
>
> We are appalled, too, by the fact that pain is most poorly managed in those most defenseless against it – the young and the elderly. Children often receive little or no treatment, even for extremely severe pain, because of the myth that they are less sensitive to pain than adults and more easily addicted to pain medication. Pain in the elderly is often dismissed as something to be expected and hence tolerated.
>
> All this needless pain and suffering impoverishes the quality of life of those afflicted and their families; it may even shorten life by impairing recovery from surgery or disease. People suffering severe or unrelenting pain become depressed. They may lose the will to live and fail to take normal health-preserving measures; some commit suicide.
>
> Part of the problem lies with health professionals who fail to administer sufficient doses of opiate drugs for pain of acute or cancerous origin. They may also be unaware of, or unskilled in using, many useful therapies and unable to select the most effective ones for specific pain conditions. Failure to understand the inevitable interplay between psychological and somatic aspects of pain sometimes causes needed treatment to be withheld because the pain is viewed as *merely "psychological."* [emphasis added][5]

The final line of this editorial eloquently buttresses the account we have given of the capture of medicine by a mechanistic and physicalistic ideology that denies reality to subjective experience. Also highly

5 Liebeskind and Melzack, "Meeting a Need for Education in Pain Management."

relevant to our subsequent discussion is the strong claim that pain is most egregiously ignored in the young and the elderly, that is, those most vulnerable and defenseless, a point I return to shortly.

The ignoring of pain is further buttressed in a 1991 paper by Ferrell and Rhiner that appeared in the *Journal of Clinical Ethics*.[6] According to the authors, although pain can be controlled effectively in 90 percent of cancer patients, it is in fact not controlled in 80 percent of such patients. A 1999 article in *Nursing Standard* shows that the Marks and Sacher problem continued into the new century. The author affirms that "more recent studies have shown that there has been little improvement over the years."[7] Note that the author is a nurse, not a doctor.

As pain was seen as medically unreal and subjective, control of pain was historically determined by strange ideological dicta even in the nineteenth century after the discovery of anesthesia. Historian Martin Pernick has illustrated this point well, by comparing hospital records on anesthetic use with ideological pronouncements and finding very high correlations.[8] For example, although affluent white women generally made up the class receiving the most anesthesia for most medical procedures, this was not true for childbirth, because it was believed that childbirth pain was divine punishment for Eve's transgression and also that women would not bond with the child unless they felt pain. Farmers, sailors, and other members of "macho" professions received very little anesthesia, as did foreigners. Black women, even when being used for painful experiments, received no anesthesia at all. Limb amputation was classified as "minor surgery." Children received more pain control because of their "innocence" (which, as we see, below, has been reversed under current ideology). Worries were expressed that anesthesia gives the doctor too much (sexual) control over the patient. The key point is that pain control even then was more a valuational and ideological decision than a strictly scientific medical one.

Thus, even with regard to anesthesia, precedent existed for odd, arbitrary, or ideologically dictated use. Inevitably, given the tendency to

[6] Ferrell and Rhiner, "High-Tech Comfort: Ethical Issues in Cancer Pain Management for the 1990s."

[7] Gray, "Do Respiratory Patients Receive Adequate Analgesia?"

[8] Pernick, *A Calculus of Suffering*.

see felt pain as scientifically less than real, and in any case unverifiable, and further given the ethics-free ideology we have discussed, a morally based dispensation of pain control was unlikely to be regnant. Indeed, we have already quoted Liebeskind and Melzack on the tendency for pain management to be minimal in the "defenseless."

There is ample evidence for this claim. In 1994 a paper appeared in the *New England Journal of Medicine* demonstrating that infants and children (who are, of course, powerless or defenseless in the above sense) receive less analgesia for the same procedures then do adults undergoing the same procedures.[9] But there is a far more egregious example of the same point in the surgical treatment of neonates – an example that never fails to elicit gasps of horror from audiences when I recount it. This was the practice of doing open heart surgery on neonates without anesthesia, but rather using "muscle relaxants" or paralytic drugs like succinylcholine and pancuronium bromide, until the late 1980s. Postsurgically, no analgesia was given.

Let us pause for a moment to explain some key concepts. Anesthesia means literally without feeling. We are all familiar with general anesthetics that put a patient to sleep for surgery. With general anesthesia, a person should feel no pain during a procedure. Similarly, local anesthetics, such as novocaine for a dental procedure, remove the feeling of pain from a particular area while procedures such as filling a tooth or sewing up a cut are performed, and the patient is conscious but does not feel pain. Though there are many qualifications to this rough and ready definition, they do not interfere with our point here.

Muscle relaxants or paralytics block transmission of nerve impulses across synapses and thus produce flaccid paralysis but not anesthesia. In other words, one can feel pain fully but one cannot move, which may indeed make pain worse, since pain is augmented by fear. First-person reports by knowledgeable physician/researchers of paralytic drugs, which paralyze the respiratory muscles so the patient is incapable of breathing on his or her own, recount the terrifying nature of the use of paralytics in conscious humans aware of what is happening. Analgesics are drugs that attenuate pain or raise patients' ability to tolerate pain. Examples are aspirin and acetominophen for headaches

9 Walco et al., "Pain, Hurt, and Harm."

or morphine, Demerol, or Vicodin. Thus, babies were receiving major open heart surgery using only paralytic drugs and experiencing countless procedures ranging from circumcision and venipuncture to frequent heel-sticks with no drugs for pain alleviation at all – neither anesthetics nor analgesics.[10]

The public became informed about the open-heart surgery in 1985, when a parent, whose own child died undergoing this sort of surgery, complained to the medical community, was essentially ignored, and went public, supported by some operating room nurses who felt strongly that babies experienced pain. The resulting public outcry caused the medical community to reexamine the practice and eventually to abolish it.[11]

The reasons that anesthesia was ignored in neonates were multiple and familiarly ideological. First, the medical community believed pain is "subjective" and thus not medically real. Second, since babies do not remember pain, pain doesn't matter. Third, it was argued and widely accepted that the neonatal cortex or other parts of the nervous system were insufficiently developed to experience pain. For example, it was said that babies' nerves were insufficiently myelinated for babies to feel pain. Fourth, since all anesthesia is selective poisoning, it was argued that anesthesia was dangerous. Many of the claims on which the objections to anesthesia were based were deftly handled in a classic paper by Anand and Hickey, entitled "Pain and Its Effects in the Human Neonate and Fetus."[12]

To the first claim that pain is (merely) subjective, the reply is simple: first, that that is equally true for adults and second, what is subjective is very real for the experiencer. (The essence of pain is that it *hurts.*) To the claim that forgotten pain doesn't matter, the simple response is that, once experienced, pain is biologically active and retards healing and is immunosuppressive even if forgotten. To this day, painful procedures such as bronchoscopy and colonoscopy are done under amnesic drugs in adults, who may feel much pain during the procedure but don't remember because of the drug. Failure to remember does not

[10] Cope, "Neonatal Pain."

[11] Circumcision Information and Resource Pages, "Pain of Circumcision and Pain Control."

[12] Anand and Hickey, "Pain and Its Effects in the Human Neonate and Fetus."

justify infliction of pain. Furthermore, babies give evidence of memory when brought back to rooms in which they underwent surgery.

Third, Anand and Hickey convincingly debunk the claim that neonates – and even preterm babies – do not feel pain. There are convincing physiological arguments that both myelination and cortical development in neonates suffice to attribute pain to infants. Behavioral changes also buttress this point.

Fourth, *all* anesthesia is dangerous, particularly when administered to sick people. The key point is that adequate anesthesia regimens exist to tilt the cost-benefit ratio in favor of using anesthesia. In a later paper (1992),[13] Anand and Hickey showed that neonates given high doses of anesthesia and analgesia for surgery fared better in terms of morbidity and mortality than children treated with light anesthesia. They demonstrated that when infants undergoing open heart surgery were deeply anesthetized and given high doses of opiates for twenty-four hours postoperatively, they had a significantly better recovery and significantly fewer postoperative deaths than a group receiving a lighter anesthetic regimen (halothane and morphine) followed post-operatively by intermittent morphine and diazepam for analgesia. The group that received deep anesthesia and profound analgesia "had a decreased incidence of sepsis, metabolic acidosis, and disseminated intravascular coagulation and fewer postoperative deaths (none of the 30 given sufentanil versus 4 of 15 given halothane plus morphine)."[14]

The conclusion of Anand and Hickey's 1987 paper is worth quoting in its entirety:

> Numerous lines of evidence suggest that even in the human fetus, pain pathways as well as cortical and subcortical centers necessary for pain perception are well developed late in gestation, and the neurochemical systems now known to be associated with pain transmission and modulation are intact and functional. Physiologic responses to painful stimuli have been well documented in neonates of various gestational ages and are reflected in hormonal, metabolic, and cardiorespiratory changes *similar to but greater than* those observed in adult subjects [emphasis added]. Other responses in newborn infants are suggestive of integrated emotional and behavioral responses to pain and are retained in memory long enough to modify subsequent behavior patterns.

[13] Anand and Hickey, "Halothane-Morphine Compared with High Dose Sufentanil for Anesthesia and Post-Operative Analgesia in Neonatal Cardiac Surgery."

[14] Ibid.

> None of the data cited herein tell us whether neonatal nociceptive activity and associated responses are experienced subjectively by the neonate as pain similar to that experienced by older children and adults. However, the evidence does show that marked nociceptive activity clearly constitutes a physiologic and perhaps even a psychological form of stress in premature or full-term neonates. Attenuation of the deleterious effects of pathologic neonatal stress responses by the use of various anesthetic techniques has now been demonstrated.... The evidence summarized in this paper provides a physiologic rationale for evaluating the risks of sedation, analgesia, local anesthesia, or general anesthesia during invasive procedures in neonates and young infants. Like persons caring for patients of other ages, those caring for neonates must evaluate the risks and benefits of using analgesic and anesthetic techniques in individual patients. However, in decisions about the use of these techniques, current knowledge suggests that humane considerations should apply as forcefully to the care of neonates and young, nonverbal infants as they do to children and adults in similar painful and stressful situations.

It is interesting to note that, as in the case of pain in animals, the scientific "reappropriation of common sense" about infant pain occurred only subsequent to public moral outrage about standard practice.

In a powerful and sensitive 1994 paper in the *New England Journal of Medicine*, Walco, Cassidy, and Schechter review some of the major arguments leading to withholding pain control from children and infants, echoing points we have seen made in Anand and Hickey.[15] These include the subjectivity of pain, the belief that children are not reliable reporters of pain, a failure to recognize individual differences in children (despite solid scientific evidence to the contrary), misinformation about the neurologic capacity to feel pain, and the "no memory" argument. Recent evidence indicates that this last point is particularly egregious, that not only does unrelieved pain disturb eating, sleeping, and arousal in the neonate, but "infants retain a memory of previous experience, and their response to a subsequent painful experience is altered"[16]; failure to control pain in infants leads to aberrant nerve growth, causing additional pain later in life.[17]

Walco et al. also raise and refute the claim that opioid analgesics cause respiratory depression or arrest. They point out that "the risk

[15] Walco et al., "Pain, Hurt, and Harm."
[16] Lee, "Managing Pain in Human Neonates."
[17] Beggs, "Postnatal Development of Pain Pathways."

of narcotic-induced respiratory depression in adults is about 0.09 percent, whereas in children it ranges between 0 percent and 1.3 percent." In most cases, the problem is solved by dose reduction, and opiate overdose can be reversed. They also indicate that fully 39 percent of physicians worry about creating addicts by use of opioids, yet this concern is baseless, with "virtually no risk of addiction associated with the administration of narcotics." Another set of arguments affirms that masking pain masks symptoms (a very common reason for not using analgesia in veterinary medicine), an absurd claim regarding major surgical postoperative pain. An additional argument affirms that "pain builds character," again an absurd argument in an infant or suffering child. As Walco et al. declare, "If there is a therapeutic benefit from a child's pain, one must be exquisitely economical with it."

The conclusion of the Walco paper is as morally sensitive and powerful as the rest:

> There are now published guidelines for the management of pain in children, which are based on recent data. However, guidelines and continuing medical education do not necessarily alter physicians' behavior. Specific administrative interventions are required. For example, hospitals may include standards for the assessment and management of pain as part of their quality-assurance programs. The Joint Commission on Accreditation of Healthcare Organizations has established standards for pain management. To meet such standards, multidisciplinary teams must develop specific treatment protocols with the goal of reducing children's pain and distress. In addition, pressure from parents and the legal community is likely to affect clinical practice.
>
> All health professionals should provide care that reflects the technological growth of the field. The assessment and treatment of pain in children are important parts of pediatric practice, and failure to provide adequate control of pain amounts to substandard and unethical medical practice.[18]

Many of the points made in the Walco paper have direct implications for other areas where pain is neglected. For example, large portions of the medical community have steadfastly opposed the use of narcotics in terminally ill patients on the dubious grounds that such patients may become addicted. The first response, of course, is, "So what if they become addicted? These drugs are cheap!" In any case, the

[18] Walco et al., "Pain, Hurt, and Harm."

medical community's ignorance in this area is appalling. Again from Walco:

> It is essential to distinguish between physical dependence (a physiologically determined state in which symptoms of withdrawal would occur if the medication were not administered) and addiction (a psychological obsession with the drug). Addiction to narcotics is rare among adults treated for disease-related pain and appears to depend more on psychosocial factors than on the disease or medically prescribed administration of narcotics. Studies of children treated for pain associated with sickle cell disease or postoperative recovery have found virtually no risk of addiction associated with the administration of narcotics. There are no known physiologic or psychological characteristics of children that make them more vulnerable to addiction than adults.[19]

We also find large numbers of physicians vehemently opposed to medical marijuana. It appears that such physicians have been swayed by simplistic government propaganda about drugs and addiction – "one shot and you're hooked." In fact, there were many regular heroin users among soldiers in Vietnam who, on their return home and no longer in stressful situations, gave up the drug and were not addicted.[20] Again, we see ideology trump both science and reason – in this case, the ideology underlying U.S. drug policy.

Another glaring example of medicine's ignoring of subjective states can be found in the history of the drug ketamine. This illustration is particularly valuable in that it demonstrates the cavalier attitude that historically obtained (and indeed still obtains) with regard to negative subjective experiences in humans and animals.

Ketamine is a cousin of phencyclidine.[21] Phencyclidine (also known as PCP) was developed in the 1950s but was found to be very dangerous in terms of hallucinations, creating violent behavior, confusion, delusions, and abusability. Various derivatives of PCP were tried until 1965, when ketamine was found to be most promising. In 1970, it was released for clinical use in humans in the United States.

[19] Ibid.
[20] Robins et al., "Drug Use in U.S. Army Enlisted Men in Vietman."
[21] MetroHealth Anesthesia, "History of Ketamine."

Ketamine was heralded as the "ideal" anesthetic, since overdose was virtually impossible, and it did not cause respiratory depression. Furthermore, it could be administered via intravenous, intramuscular, oral, rectal, or nasal routes. Ketamine is profoundly analgesic (pain relieving) for somatic or body pain, though it is of no use for visceral (gut) pain. It has been particularly useful in human medicine for treating burn patients and changing dressings.

In the 1980s, while researching a paper on pain, I looked at ketamine in some detail.[22] In the first place, I found that it was used very frequently in research as a sole surgical anesthetic in small rodents and other animals and in veterinary practice as a standard "anesthetic" for spays. Since it is emphatically not viscerally analgesic, this meant that in such procedures it was being used as a restraint drug. Under ketamine, animals are "disassociated" – experience a strong feeling of disassociation from the environment and are immobilized in terms of voluntary movement. When I watched a visceral surgery on a cat done with ketamine, I could see obvious signs of pain when the viscera were cut or manipulated. In essence, this means that when ketamine alone was used for visceral procedures, the animals felt pain but were immobilized.

In human medicine, ketamine was used for a wide variety of somatic procedures, such as burn dressing change and plastic surgery. But, by 1973, the medical community had become aware of the fact that ketamine was capable of engendering significantly "bad trips" in a certain percentage of patients, though many experienced pleasant hallucinations. Contributing to this awareness was a letter published in *Anesthesiology* in 1973 by Robert Johnstone, wherein the contributor, an M.D. anesthesiologist, graphically described his experiences under ketamine as a research subject.[23] It is important to stress that Dr. Johnstone had taken several different narcotics and sedatives before the ketamine experience and had had no problems. The ketamine experience, however, was quite different:

> I have given ketamine anesthesia and observed untoward psychic reactions, but was not concerned about this possibility when the study began. After my experience, I dropped out of the study, which called for two more exposures

[22] Rollin, "Pain, Paradox, and Value."

[23] Johnstone, "A Ketamine Trip."

of ketamine. In the several weeks since my ketamine trip, I have experienced no flashbacks or bad dreams. Still I am afraid of ketamine. I doubt I will ever take it again because I fear permanent psychologic damage. Nor will I give ketamine to a patient as his sole anesthetic agent.[24]

Here is Johnstone's description of what occurred:

My first memory is of colors. I saw red everywhere, then a yellow square on the left grew and crowded out the red. My vision faded, to be replaced by a black and white checkerboard which zoomed to and from me. More patterns appeared and faded, always in focus, with distinct edges and bright colors.

Gradually I realized my mind existed and could think. I wondered, "What am I?" and "Where am I?" I had no consciousness of existing in a body; I was a mind suspended in space. At times I was at the center of the earth in Ohio (my former home), on a spaceship or in a small brightly-colored room without doors or window. I had no control over where my mind floated. Periods of thinking alternated with pure color hallucinations.

Then I remembered the drug study and reasoned something had gone wrong. I remembered a story about a man who was awake during a resuscitation and lived to describe his experience. "Am I dying or already dead?" I was not afraid, I was more curious. "This is death. I am a soul, and I am going to wherever souls go." During this period I was observed to sit up, stare and then lie down.

"Don't leak around the mouthpiece!" were the first real sounds I heard. I couldn't respond because I didn't have a body. Thus began my cycling into and out of awareness – a frightening experience. I perceived the laboratory as the intensive care unit; this meant something had gone wrong. I wanted to know how bad things were. I now realized I wasn't thinking properly. I recognized voices, then I recognized people. I saw some people who weren't really there. I heard people talking, but could not understand them. The only sentence I remember is "Are you all right?" Observers reported a panicked look and defensive thrashing of my arms. I screamed "They're after me!" and "They're going to get me!" I don't recall this or remember the reassurances given me.

I then became aware of my body. My right arm seemed withered and my left very long. I could not focus my eyes. Observers reported marked nystagmus. I recognized the ceiling, but thought it was covered with worms (apparently cued by the irregular depressions in the soundproof blocks). I desperately wanted to know what was reality and to be part of it. I seemed to be thinking at a normal rate, but couldn't determine my circumstances. I couldn't speak or communicate, but once, recognizing a friend next to me, I hugged him until I faded back to abstractness.

[24] Ibid., p. 461.

> The investigators gave me diazepam, 20 mg, and thiopental, 150 mg, intravenously because I was obviously anxious, and I fell asleep. When I awoke it was five hours since I had received ketamine. I promptly vomited bilious liquid. Although I could focus accurately, I walked unsteadily to the bathroom. I assured everyone "I'm OK now." Suddenly I cried with tears for no reason. I knew I was crying but could not control myself. I fell asleep again for several hours. When I awoke I talked rationally, was emotionally stable and felt hungry. The next day I had a headache and felt weak, similar to the hangover from alcohol, but functioned normally.[25]

Today, of course, ketamine (known by the street name "special K") is classified as a schedule III drug, not only because it is widely abused, but because it has become a rape drug, in virtue of the immobility and "paralysis of will" it produces. And there are countless examples in literature of vivid depictions of bad ketamine trips going back to 1973.[26] An additional troubling dimension of ketamine use became known at this time: the tendency of ketamine to produce unpredictable "flashbacks," much in the manner of LSD.

When researching all this in 1985, my main interest was its use in animals. I therefore approached some world-renowned veterinary anesthesia colleagues, who confirmed, first of all, its misuse for visceral surgery. I then asked about "bad trips." There was no literature on this, I was told, but anecdotally, such occurrences were obvious. As my colleague put it, "Most animals (cats) see little pink mice; but some see giant, ferocious pink rats." Despite this observation, I have never seen any discussion in the veterinary literature of "bad trips." Similarly, there is no literature on deviant behavior indicating possible flashbacks in animals, but I have been told of owners reporting complete personality changes in animals after ketamine dosing, one woman claiming that the hospital had given her back the wrong animal! The failure of veterinary medicine to even discuss such potential problems eloquently attests to the perceived irrelevance of bad subjective animal experiences to scientific veterinary medicine.

Continuing my research on ketamine in 1985–86, I was curious about how ketamine use had changed since the 1973 revelations of bad trips and flashbacks. Much to my amazement, I now found that

[25] Ibid.

[26] See cases in EROWID Experience Vaults on the Web, e.g., Abe Cubbage, "A Trend of Bad Trips."

ketamine was largely being used "on the very young (children) and the very old (the elderly)."

For the next few months I searched the anesthesia literature, journals and textbooks, to find out what unique physiological traits were common to the very young and the very old that made ketamine a viable drug at these extremes, but not for people in the middle. I got nowhere. Finally, by sheer coincidence, I happened to be at a party with a human anesthesiologist and asked him about the differing physiologies. He burst out laughing! "Physiology?" he intoned. "The use has nothing to do with physiology. It's just that the old and the young can't sue and have no power!" In other words, their bad subjective experiences don't matter.

This was confirmed for me by one of my students, who had a rare disease since birth that was treated at a major research center. He told me that procedures were done under ketamine, which he loathed in virtue of "bad trips," until he turned sixteen, at which time he was told, "Ketamine won't work anymore."

If there ever was a beautiful illustration of ideological, amoral, cynical denial of medical relevance of subjective experience in human and veterinary medicine, it is the above account of ketamine. Unfortunately, there is more to relate on this issue. I now discuss the International Association for the Study of Pain (IASP) definition of pain that was widely disseminated until finally being revised in 2001 to mitigate some of its absurdity.

IASP is the world's largest and most influential organization devoted to the study of pain. Yet the official definition of pain entailed that infants, animals, and nonlinguistic humans did not feel pain! In 1998, I was asked to criticize the official definition of pain, which I felt was morally outrageous in its exclusion of the above from feeling pain and in reinforcing the ideological denial of subjective experience to a large number of beings to whom we had moral obligations. Leaving such a definition to affirm the scientific community's stance on felt pain was a matter causing both moral mischief and ultimately a loss of scientific credibility. The discussion that follows is drawn from my remarks at the IASP convention of 1998 and my subsequent essay version published in *Pain Forum.*[27]

[27] Rollin, "Some Conceptual and Ethical Concerns about Current Views of Pain."

It is a major irony that although the definition of pain adopted by the IASP was cast into its current form for laudable moral reasons, it has given succor to neo-Cartesian tendencies in science and medicine, and in fact has the potential for supporting morally problematic behavior. Dr. Harold Merskey, a principle architect of the definition, has explained at the 1998 American Pain Society meeting in San Diego that the initial definition of pain as "an unpleasant sensory and emotional experience associated with actual or potential tissue damage, or described in terms of such damage"[28] was later modified in a note to allow for the reality of pain in adult humans where there was no organic cause for the pain and no evident tissue damage. The note affirmed:

> Pain is always subjective. Each individual learns the application of the word through experiences related to injury in early life.[29]

It continues:

> Many people report pain in the absence of tissue damage or any likely pathophysiological cause: Usually this happens for psychological reasons. There is usually no way to distinguish their experience from that due to tissue damage if we take the subjective report. If they regard their experience as pain and if they report it in the same ways as pain caused by tissue damage, it should be accepted as pain.[30]

In other words, linguistic self-reports of pain should be accepted as proof of the existence of genuine pain in linguistically competent beings, a move designed to encourage medical attention to pain even in the absence of a proximate stimulus involving tissue damage. This was clearly a praiseworthy, morally motivated move, which also spurred research into areas such as chronic pain that might have been ignored in the absence of a definition stressing the subjective side of pain and its linguistic articulation.

Unfortunately, however, the definition's emphasis on the connection between pain and full linguistic competence have led to a neo-Cartesian tendency to make such linguistic competence a necessary and sufficient condition for attributing pain to a being (Descartes had

[28] International Association for the Study of Pain (IASP), "Pain Terms."
[29] Merskey and Bogduk, *Classification of Chronic Pain.*
[30] Ibid.

famously argued that only creatures with language could be said to possess mind). "Mere" behavior does not license the confident or certain attribution of pain to an organism because only words describe the subjective. Merskey says: "The behavior mentioned in the definition is behavior that describes the subjective state and that is how matters should remain."[31] Merskey also stated:

> The very words "pain behavior" are often employed as a means to distinguish between external responses and the subjective condition. I am in sympathy with Anand and Craig in their wish to recognize that such types of behaviour are likely to indicate the presence of a subjective experience, but the behavior cannot be incorporated sensibly in the definition of a subjective event.[32]

Despite Merskey's own professed belief that "there is an almost overwhelming probability that some speechless organisms suffer pain, including neonates, infants and adults with dementia"[33] (he does not mention animals), he nonetheless classifies such pain as "probable" or "inferred," in contradistinction to the certainty accompanying claims by linguistic beings. As Anand and Craig[34] and Craig[35] have argued, this ultimately draws a major ontological and epistemological gulf between linguistic and nonlinguistic beings in relation to the presence and certainty of experienced pain. This, in turn, helps to justify the well-documented tendency of researchers and clinicians to undertreat or fail to treat altogether pain in neonates, infants, young children, and animals, all of whom lack full linguistic ability.

It is thus disturbing to find a neo-Cartesian element infiltrating these recent discussions of pain, suggesting that only linguistic beings are capable of experiencing pain as something of which they are aware and that only verbal reports allow us to "really" know that a being is in pain. Aside from the ethical damage that such a view can create by implying that animals, neonates, and prelinguistic infants do not "really feel pain," promulgation of this view is dangerous to the

[31] Merskey, "Pain, Consciousness and Behavior."
[32] Merskey, "Response to Editorial."
[33] Merskey, "Pain, Consiousness and Behavior."
[34] Anand and Craig, "New Perspectives on the Definition of Pain."
[35] Craig, "Implications of Concepts of Consciousness for Understanding Pain Behavior and the Definition of Pain."

methodological assumptions underlying science, as well as to scientific credibility in society in general.

To illustrate the methodological pitfalls inherent in such a view, consider the thesis once raised by Bertrand Russell: "How do we know that the world was not in fact created 10 seconds ago, complete with fossils, etc. and us with all of our memories?"[36] Or better yet, consider the following critique of the very possibility of science: "Look here. Science claims to give us explanations of phenomena that take place in the physical world we all share. Yet, in point of fact, our only access to the real physical world is through our experiences, our perceptions, which are totally *subjective*, unique to each perceiver. After all, it is notorious that I can't know what you perceive. You may not see red as I do, or hear sounds as I do. How then, do we ever get to an 'objective' world by summing a whole bunch of inherently subjective perceptions and experiences?"

This, of course, is an argument for solipsism that few if any scientists worry about when they attempt to explain the nature and causes of disease, earthquakes, atomic and subatomic phenomena, mind, and so forth. Why don't they worry about it? Because both of the concerns detailed above are, like the existence of God or of an immaterial soul, ultimately *metaphysical* hypotheses, which gathering data or doing experiments can never refute or confirm, and scientific activity is archetypically tethered, however indirectly, to what can be confirmed by observation and experiment.

Why go off on this tangent? Very simply, because the thesis that only linguistic beings can *feel pain* or be aware of pain or give us evidence that they have pain is precisely such a metaphysical thesis as well, to which no amount of data is ultimately relevant.

We all recognize that when we judge that another person feels pain, we are making a fallible claim. The person (i.e., the linguistic being) may be malingering, faking, or acting. So we seek other evidence, such as signs of injury or knowledge that the injury or condition in question produces pain generally; we check our ability to lessen the pain with anesthetics or analgesics; we look for involuntary moans and groans that the person may emit when fully or partially asleep; and so on. Like all empirical claims, judgments that someone is in pain are in

[36] Russell, *Problems of Philosophy*.

principle falsifiable. The presence of language is certainly not definitive, as language can be used to mislead and befuddle, as well as to inform. In practicing science or medicine, we go with the weight of evidence: the presence of inflammation, guarding of a limb, change in pallor, reluctance to eat, and so forth. No scientist of any credibility would affirm that he or she is withholding judgment that another person feels pain in such circumstances just because the scientist cannot, in principle, feel the same feeling the other person experiences, or perhaps does not at all. In good scientific fashion, one goes with the weight of evidence, not with skepticism based in untestable metaphysical possibilities. Doing the latter would be exactly like a scientist rejecting another scientist's experimental data solely on the grounds of the metaphysical claim that he or she cannot be sure that the other person perceives at all, or perceives as we do because we cannot experience their perception.

If science proceeds, then, by weight of empirical evidence in general and in the attribution of felt pain in particular, and does not allow theses based in metaphysical possibilities of solipsism (lack of absolute certainty that anyone else perceives as I do, etc.), then it is equally a major logical error to deny felt pain to nonlinguistic beings a priori, regardless of what physiological, behavioral, factual, or theoretical evidence exists to vouch for such felt pain.

Certainly, that evidence is abundant in our experience with animals, so much so that ordinary experience, common sense, and language do not infer (or reason) that an animal is in pain but perceive it immediately. If a dog is run over by a car, is not unconscious, has a compound fracture jutting out of his skin and a crushed limb, is howling and whining and shivering, we automatically assume he is in pain. If someone asks, "But how do you know?" we assume that he is either demented or making a bad joke.

No one knows better than pain scientists that this powerful, unshakeable, common-sense response is strongly buttressed by myriad scientific evidence, such as the following: that animal pain physiology and neuroanatomy is essentially the same as human down the phylogenetic scale; that pain biochemistry – including the emergence of endogenous opiates after trauma and the presence of such chemicals as bradykinin and substance P in painful areas – is similarly phylogenetically continuous; that pain behavior and signs of pain, while certainly different in some marked ways across species, is no more different from what it is across human cultures and subcultures and is very similar in many ways (punch a Doberman pinscher, a tiger, a buffalo, a shark, and a gangbanger in the mouth and see the reaction, if you still doubt me; recall the guarding of limbs across species, etc.); that Darwinian evolutionary continuity makes the emergence of felt pain in humans alone highly suspect, especially given the above-mentioned similarities; that if animals did not feel pain, they could not serve as pain models for humans in pain and analgesia studies; that anesthetics and analgesics seem to have the same beneficial effects on animals as in humans, from quieting signs of suffering to accelerating healing; and that preemptive analgesia works the same in humans and (at least) in mammals.

Indeed, let us recall that one eminent pain physiologist, the late Dr. Ralph Kitchell, co-editor of the American Physiological Society symposium volume on animal pain, has argued for the possibility that animals, in general, *feel* pain more acutely than humans. According to Kitchell, response to pain is divided into a sensory-discriminative dimension and a motivational-affective dimension. The former is concerned with locating and understanding the source of pain, its intensity, and the danger with which it is correlated; the latter with escaping from the painful stimulus. Kitchell speculates that since animals are more limited than humans in the first dimension because they lack human intellectual abilities, it is plausible to think that the second dimension is correlatively stronger, as a compensatory mechanism. In short, since animals cannot deal intellectually with danger and injury as we do, their motivation to flee must be correlatively stronger than ours – in a word, they probably hurt more. [37]

There is no question in my mind that what we call language is unique to humans, and, very speculatively, that something approximating language belongs to few other species, perhaps great apes and dolphins. That does not mean, however, as Descartes and our current Cartesians conclude, that language is the only sure way we can know that a being is in pain, fear, anxiety, distress, joy, sexual excitement, and other fundamental and basic modes of awareness. I have

[37] Kitchell and Guinan, "Nature of Pain in Animals."

argued elsewhere against the traditional belief that there is a clear and unbridgeable gulf between the sort of meaning we find in natural signs (e.g., clouds mean rain or smoke means fire) and the sort of meaning we find in conventional (or man-made) signs, such as the word "cloud" in English means, "visible condensed water droplets." As philosopher George Berkeley affirmed, nature is full of meaning, and science can be viewed as, in his metaphor, learning to read the language of nature. Animals, although presumably lacking language, find meaning in the world (e.g., a scent meaning prey) and also impart meaning to other animals and humans (e.g., threats).

It is possible to suggest that a being with language can communicate better about the nature of pain than one lacking language, but even if this is true, that does not mean that one cannot communicate the presence and intensity of pain without language, by natural signs. Recall that language does not help us much in describing our pain to others; verbal reports are notoriously unreliable. Recall too that in addition to helping us to communicate, language helps us to prevaricate and conceal. The posture and whimpering of an injured animal or the groans of an injured person are, in my mind, far more reliable indicators of the presence and intensity of pain than are mere verbal reports. Let us further recall that the natural signs we share with animals are far more eloquent and persuasive signs of primordial states of consciousness such as love, lust, fear, and pain than is Shakespearean English – words fail in the most fundamental and critical areas (as when a physician asks you to describe your pain).

It has sometimes been suggested that the possession of linguistic concepts is related to pain in the following way: Only a being with language, and the temporal concepts provided by language, can project ahead into the future or backward into the past. Much of our pain is associated with such projection – the pain of a visit to the dentist is surely intensified by the magnified recollections of previous pain, filtered through imaginative anticipations of horrific future scenarios informed by having seen the movie *Marathon Man*, wherein one's dentist turns out to be (literally) a Nazi war criminal. In the absence of concepts of past and future, animals cannot recollect or anticipate, being, as it were, stuck in the now. Thus, the claim is that their pain is considerably more trivial than ours.

Aside from the obvious objections – if animals have no access to the past, how can they learn (which they clearly do) and if animals have no concept of the future, how can a dog beg for food or a cat wait patiently for a mouse (which they clearly do) – there is a much more profound issue raised by this argument. If animals are indeed inexorably locked into what is happening in the here and now, as the above argument suggests, we are all the more obliged to try to relieve their suffering, because they themselves cannot look forward to or anticipate its cessation, or even remember, however dimly, its absence. If they are in pain, their whole universe is pain; there is no horizon; they are their pain. So, if this argument is indeed correct, then animal pain is terrible to contemplate, for the dark universe of animals logically cannot tolerate any glimmer of hope within its borders.

In less dramatic and more philosophical terms, Spinoza pointed out that understanding the cause of an unpleasant sensation diminished its severity and that, by the same token, not understanding its cause can increase its severity.[38] Common sense readily supports this conjecture; indeed, this is something we have all experienced with lumps, bumps, headaches, and, most famously, suspected heart attacks that turn out to be gas pains.

Spinoza's conjecture is thus borne out by common experience and by more formal research. But this would be reason to believe that animals, especially laboratory animals, suffer more severely than humans, since they have no grasp of the cause of their pain, and thus, even if they can anticipate some things, then have no ability to anticipate the cessation of pain experiences outside their normal experience.

We further know that humans who cannot feel pain, even though they have the full nociceptive machinery, do not fare well as far as survival is concerned. Whether the inability to feel pain is a genetic anomaly or a result of diseases such as Hansen's disease or diabetes, such human lose limbs, contract infection, and have truncated lives. Is it really plausible to suggest that all animals without language are permanently in that state? And if they are, how do they thrive?

One final argument against making the possession of language a necessary condition for feeling pain is the following. Philosopher

[38] Spinoza, *Ethics* (Parts 3–5).

Thomas Reid pointed out, quite reasonably, that since babies are not born linguistic beings, they must acquire it.[39] Even if Chomsky is correct that the skeleton for language is innate, it must still be actualized by experience of some language. This in turn entails that people must be capable of experience before they have language, else they could not learn it (or actualize their innate capacity for it). But if nonlinguistic (or prelinguistic) experience is possible, surely one of the most plausible candidates for such experience is pain, first of all because it is so essential to survival, and second, because we have so much evidence (discussed earlier) that nonlinguistic beings in fact experience pain.

For all of these reasons, then, including linguistic ability in the requirements for feeling pain or attributing pain to another represents a combination of bad science and bad philosophy.

I went on to argue that this definition leads to bad ethics among scientists in ignoring treatment of pain in nonlinguistic beings, and also a bad picture of science to society, something very undesirable at a historical moment wherein society has lost the old utopian confidence in science and scientists. Presumably, some sense of the moral/political climate drove IASP to modify this definition in 2001 in a minimalistic way. In a note, the definition now affirms:

> The inability to communicate verbally does not negate the *possibility* [emphasis added] that an individual is experiencing pain and is in need of appropriate pain-relieving treatment.[40]

This sounds far more like a concession to political reality than the embracing of a major conceptual upheaval.

In any event, the attitude exhibited in the IASP definition is perfectly consonant with what we have documented about human pain, and, given the situation with human pain, the reader can guess how cavalierly animal pain was treated.

Indeed, for younger people trained before the late 1980s, it is difficult to fathom the degree to which the denial of consciousness, particularly animal consciousness and particularly pain, was ubiquitous in science. In 1973, the first U.S. textbook of veterinary anesthesia was

[39] Reid, *Inquiry into the Human Mind on the Principles of Common Sense*, chapter 5.
[40] IASP, "IASP Pain Terminology," pain.

published, by Lumb and Jones.[41] Although the book gave numerous reasons for anesthesia (to keep the animal from hurting you, to keep it from injuring itself, to allow you to position the limbs for surgery), the control of felt pain was never even mentioned. When I went before Congress in 1982 to defend our laboratory animal legislation, I was advised to demonstrate that such laws were needed. To accomplish this goal, I did a literature search on laboratory animal analgesia and, *mirabile dictu*, found only one or two references, one of which argued that there *should be* such knowledge.

In 1983, the crescendo of concern among the public about animal pain was so great that the scientific community felt compelled to reassure the public that animal pain was indeed an object of study and concern, so they orchestrated a conference on pain and later published a volume entitled *Animal Pain: Perception and Alleviation*.[42] Despite the putative purpose of the volume, virtually none of the book was devoted to perception or alleviation of felt pain. As a result of scientific ideology, pain was confused with nociception, so that the volume focused on the neurophysiology and electrochemistry of pain, what I at the time called the "plumbing of pain," rather than the morally relevant component of pain, namely, that it *hurts*.

Most surprising to members of the general public is the fact that veterinarians were as ignorant and skeptical about animal consciousness, even animal pain, as any scientist. To this day, and certainly in the 1980s, veterinarians called anesthesia "chemical restraint" or "sedation" and performed many procedures, for example, horse castration, using physical restraint, what was jocularly called "bruticaine," or using paralytic drugs such as succinylcholine chloride, which is a curariform drug inducing flaccid paralysis, not anesthesia. Indeed, one veterinary surgeon told me that, until he taught with me, it never dawned on him that the horse being castrated under succinyl hurt!

This sort of absurdity also occurred in physiological psychology. I have already mentioned the psychological community's rejection of animal consciousness. Yet the same community regularly performed stereotaxic brain surgery and brain stimulation using succinylcholine

[41] Lumb and Jones, *Veterinary Anesthesia.*

[42] Ramey and Rollin, *Complementary and Alternative Veterinary Medicine Considered.*

without anesthesia, because the psychologists wanted the animals "conscious."

The fact that ideology could trump logic and even reason was manifest in this area. In the late 1970s, I debated a prominent pain physiologist. His talk expounded the thesis that since the electrochemical activity in the cerebral cortex of the dog (his research model for studying pain) was different from such activity in the human, and since the cortex was the seat of processing information, the dog did not (really) feel pain the way humans did. His talk took an hour, and I was expected to rebut his argument. My rebuttal was the shortest public statement I ever made. I said, "As a prominent pain physiologist, you do your work on dogs. You extrapolate the results to people, correct?" "Yes," he said. "Excellent," I said. "Then either your speech is false or your life's work is!"

In a similar vein, I experienced the following incident. In the mid-1980s, I was having dinner with a group of senior veterinary scientists, and the conversation turned to the subject of this chapter, namely, scientific ideology's disavowal of our ability to talk meaningfully about animal consciousness, thought, and awareness. One man, a famous dairy scientist, became quite heated. "It's absurd to deny animal consciousness," he exclaimed loudly. "My dog thinks, makes decisions and plans, etc., etc.," all of which he proceeded to exemplify with the kind of anecdotes we all invoke in such common-sense discussions. When he finally stopped, I turned to him and asked. "How about your dairy cows?" "Beg pardon?" he said. "Your dairy cows," I repeated, "do they have conscious awareness and thought?" "Of course not," he snapped, then proceeded to redden as he realized the clash between ideology and common sense, and what a strange universe this would be if the only conscious beings were humans and dogs, perhaps humans and *his* dog.

A colleague of mine, working on her doctorate in the mid-1980s in anesthesiology, was studying anesthesia in horses. The project involved subjecting the animal to painful stimuli and seeing which drugs best controlled the pain response. When she wrote up her results, her committee did not allow her to say that she "hurt" the animals nor could she say that the drugs controlled the pain – that was ideologically forbidden. She was compelled to say that she subjected them to a stimulus and to describe how the drugs changed the response.

There were many rationalizations ingrained in researchers and veterinarians buttressing the formidable ideological denial of pain in animals. For example, it was dogma among surgeons that the postsurgical whimpering, shivering, and crying that I saw as indicative of pain in postsurgical animals was not pain at all, but after-effects of anesthesia, as mentioned above. When pressing for analgesia, I was told that the pain was necessary to keep animals still after surgery or injury. In actual fact, of course, animals are smart enough to avoid exertion when sick or injured. It is humans who keep working or playing. Furthermore, as we know from our own experiences of analgesia, it does not eliminate pain; rather, it raises our pain-tolerance threshold, so that we do not suffer as much.

Still others affirmed that cattle did not need postsurgical pain control because they "eat right after surgery," and thus could not possibly be in pain. Because of such stoic behavior, some veterinarians still spay (and, of course, castrate) cattle with no anesthesia. The answer, of course, is that stoic behavior does not prove that the animals are not feeling pain. We need to recall that cattle are a prey species, and in nature, herds of cattle are always accompanied by circling predators, ever-vigilant to signs of weakness or debilitation. Any cow that did not therefore behave normally when in pain would be quickly culled by predation.

It was claimed that dogs did not hurt after abdominal surgery for anatomical reasons – their viscera are suspended in a mesenteric sling. Similarly, one heard that dogs were alert and wagging their tails postsurgically, so they surely did not hurt. Various researchers have done much to dispel such myths. Veterinarian Bernie Hansen has regularly pointed out that the presence of humans in a postsurgical ward significantly skews the animals' behavior and that one sees a different story when one videotapes the animals in the absence of humans. In one particularly dramatic tape, Hansen shows a Malamute dog who had experienced major disk surgery, yet in the presence of people, sat up putatively bright and alert and never even lay down to rest. The taping provided dramatic new evidence. When people were not present, the dog would involuntarily start to sink into a sleeping position. But his back hurt so much that any attempt to lie down would awaken him, as evidenced by a pathetic series of cries and whimpers!

As mentioned earlier, the federal legislation did much to eliminate agnosticism about, and denial of, felt pain in research animals and to force its use even by those who remain agnostic, federal law being one of the few levers powerful enough to overturn ideology. Papers on analgesia and pain have proliferated, and in general the analgesia requirements are quite well enforced by animal care and use committees. Since most veterinarians in academe do research, they have communicated the need for and methods of pain control to their students, who in turn take this knowledge, plus the socio-ethical imperative to control pain, into their jobs after they graduate.

Equally important, with the extraordinary recognition of the emotional role that companion animals play in people's lives, public demand for pain control for their animals has become loud and forceful. This not only has forced veterinarians in practice to set aside their denial of pain, it has again led to increased academic attention to pain control. Once again, social ethics drives transcendence of ideology.

Drug companies were quick to acknowledge the suddenly huge market for pain control in animals. Particularly relevant to our discussion is the story of Pfizer and Rimadyl®. Rimadyl® is the trade name for carprofen, a nonsteroidal anti-inflammatory drug used for analgesia in skeleto-muscular problems. Originally developed as a human drug, carprofen showed no great advantages over other anti-inflammatories, and thus was not marketed. However, someone at Pfizer thought of trying it on dogs, where it gave spectacular pain control results. Pfizer began a successful advertising campaign showing older dogs unable to romp because of skeleto-muscular pain and affirming that Rimadyl® could control such pain. Shortly thereafter, I was approached by Pfizer representatives, who told me that the biggest obstacle to marketing the drug was veterinarians, whose ideology-based denial of animal felt pain prevented them from prescribing pain control. In what was surely a first for a philosopher, I worked with Pfizer to help to lay bare these ideological presuppositions and overcome them. In the end, Rimadyl®, it is rumored, sold close to $1 billion in one year soon after being marketed.

One can argue that, in terms of rate of change, the control of animal pain probably proliferated far more rapidly than what we have indicated about human pain. I largely credit the federal law and the animal research community, who "recollected" common sense and common decency about animal pain when faced with the law. With no law driving control of pain in babies, children, or the disenfranchised, change has been slower. On the other hand, aggressive social concern about pain has created increasing amounts of attention to it. Both human and veterinary medicine now specifically address pain management in the process of accrediting hospitals.

One need only look at the over-the-counter medication for sale in any pharmacy to realize that, at the present, we are not a culture that makes a virtue of stoic enduring of pain and suffering. This is not only true of physical pain – commercials relentlessly press mood-altering drugs, male erectile enhancers, and cures for the "heartbreak of toenail fungus" (I am not making this up!). The Spartan ideal of stolid acquiescence while a fox disembowels you is a source of amazement in all but farm kids, athletes, and some military professionals. Pectoral implants and calf implants, simulating musculature in men without the hard work, are among the fastest growing procedures in plastic surgery.

Though I cannot prove this claim, it seems fairly evident that the neglect of felt pain in human and veterinary medicine has drawn people to alternative medicine. Alternative practitioners, if nothing else, are generally highly sympathetic and empathetic. Whether their treatment modalities work or not, they project care and concern, which people sometimes forget is not a substitute for effective treatment. Purveyors of effective treatment, possessed by scientific ideology, may be guilty of lack of empathy and may focus only on the disease. The result is an extraordinary groundswell of support for alternative medicine, including modalities that have been shown not to work or cannot possibly work (e.g., homeopathy), if modern science is true.

This may appear unintelligible to scientists – after all, how can people opt for what doesn't work over what works? The key point, though, is to remember what people mean by "what works." When ordinary people say assuredly that someone is a "good doctor" or a "good vet," they do not mean that they have studied the practitioners' cure rate or educational credentials – they mean that they are empathetic and seem to *care*. Thus it is obviously not enough for scientific medicine to do well on double-blind clinical trials. It must also meet socio-ethical demand for empathy and control of pain and suffering. And this, of course, means that it must abandon scientific ideology's ignoring or bracketing subjective states as irrelevant. It does not mean, in my opinion, trying uneasily to coexist with nonevidence-based alternative medicine.[43]

One final issue needs to be discussed. It will be recalled that the federal laboratory animal laws that forced what we have called the "reappropriation of common sense" on pain also contained a proviso mandating control of "distress." Beginning in 1985, however, the USDA, in writing regulations interpreting the act, focused exclusively on pain, thereby upsetting many activists. In my view, this was extremely wise, though not necessarily intentionally so. The point is that now USDA has begun to look at distress, but is doing so after the acceptance of pain has become axiomatic. Had they demanded control of pain and distress from the beginning, little progress would have been made on either category, the dual task appearing far too formidable to allow for progress on either front.

The situation is quite different now. I recently attended a conference of experts on how to deal with "distress." The preliminary discussion illuminated a fascinating *leitmotif* common to many participants. "While pain is tangible, real easy to get hold of," the argument went, "distress is far more amorphous and opaque." I was genuinely amused by this, and altered my keynote address to acknowledge the source of my amusement. "Almost twenty-five years ago exactly," I said, "I attended a very similar conference on pain, sponsored by the same people. At that time, I argued that animals felt pain, and that that pain could be known to us and controlled. An NIH official was there and said nothing, but called the Dean of my school to tell him that I was a viper in the bosom of biomedicine, and students should not be exposed to my ideas! The point is that felt pain was as remote and outlandish to scientific ideology then as distress seems to be today." I also pointed out that if 500 million dollars were made available for distress research, it would not go begging and unclaimed. The distress issue,

[43] Kitchell and Erickson *Animal Pain*; R. L. Kitchell and R. D. Johnson, "Assessment of Pain in Animals," in G. Moberg (ed.), *Animal Stress* (Bethesda, Md.: American Physiological Society, 1985).

too, is simply reappropriating ordinary common sense on negative mental states or emotions in animals, such as fear, boredom, loneliness, social isolation, or anxiety, and then providing science-based clarification of these concepts and their operational meaning and criteria for identifying them. I am certain that in twenty-five years, in retrospect, distress will look as transparent as pain.

10

Ethics in Science

In the course of reflecting on the issues dealt with in this book, I was reminded of an incident that took place when I was in the fifth or sixth grade during my elementary education. One of my elementary school's graduates had gone on to Harvard to do a Ph.D. with David Riesman, the renowned sociologist. For his Ph.D., he was studying how various groups in society viewed scientists. Toward that end, he administered a questionnaire to all the students in my grade asking us a variety of questions, essentially aimed at detecting whether children viewed scientists as markedly different from other people. The only question I remember was the following: What is a scientist most likely to do while on vacation – study a new science, work on research, or do what other people do? I recall thinking, "Well, scientists are human, just like everyone else, so they do what all people do – spend time with family, travel, etc." The graduate student called me out of class a week later, after processing the data, and informed me that I had the best understanding of how scientists behave and asked me to further elaborate on my answers. I remember explaining that people's humanness took precedence over their occupations – how could it not?

Years later, in reading Hume, I recalled my opinion, and had it buttressed when Hume said, in essence, that one can be a philosopher (or scientist) but one must be first a man.[1] As skeptical as he might have been as a philosopher about mind, body, causality, continuity of

[1] Hume, *A Treatise of Human Nature.*

personhood, and so on, he points out that common sense will not allow him to maintain the skeptical position. Rather, as soon as he hears laughter or someone pouring wine, he is thrust back into a mode of natural belief. This would of course be equally true of a scientist. However much a physicist may rationally believe that the world is nothing but microparticles in motion, he cannot help seeing the woman he loves as more than particles or hearing a symphony as more than sound waves propagated through air.

It was this sort of insight that led me to the unstartling conclusion that scientists are human, with the full human complement of warts, wens, biases, foibles, and peccadilloes. This in turn, brought me to some healthy skepticism about scientific ideology when I was studying science. Though I parroted the claims I was taught about value-free science, I had trouble accepting the reality of an objective and directive scientific method, and even more trouble swallowing the material I learned about science being independent of culture and history and, above all else, being "objective." In fact, when studying positivist philosophy of science, I secretly referred to it as "philosophy of science fiction," more a matter of philosophical ingenious invention than an X-ray of actual scientific practice. Taking a graduate seminar on methodology of social inquiry helped to expel the "value-free" beliefs, as I learned that socio-ethical commitments determined what were permissible conclusions to studies. My full abandonment of scientific ideology occurred when I began to work with scientists on a daily basis and realized how value-driven, culturally faddish, and political was science funding and even scientific methodology, as when only males were studied in clinical studies, or when IQ was banished for reasons of political correctness, or when my biomedical scientist colleagues had to spin their research programs every few years to make them relevant to AIDS or whatever happened to be trendy.

Nonetheless, as we saw earlier, scientific ideology endured and thrived in many modalities we have discussed. In what I consider to be far and away the best book ever done on fraud and deception in science, William Broad and Nicholas Wade produced *Betrayers of the Truth,* which laid bare additional ideological components I had failed to note. In the following discussion, my debt to both Wade and Broad and to my twenty-eight years of experience working with biomedical scientists on a daily basis should be evident. I proceed by first mentioning additional scientific ideology components and showing how they do not fit day-to-day scientific practice. Many of the ideological components are at best ideals to aspire to rather than canons governing actual scientific behavior.

One noteworthy component of the ideology is that science (or, more accurately, scientists) are wholly devoted to truth-seeking, which helps them to transcend ordinary, petty, human motivations. This view of scientists or, for that matter, of philosophers as being something of a priesthood wholly serving truth is quite old, reaching into antiquity. Thales, arguably the first philosopher/scientist, seeking to know the nature of things, and concluding "All is water," is depicted as being so absorbed in seeking knowledge and truth that he fell into a well while gazing at the heavens! Less well known is the story of how he deployed his scientific knowledge to predict a bad olive harvest, cornered the market on olives that year, and established himself as independently wealthy! In other words, savants of all sorts are still human – Hume admits to being almost consumed by the desire for fame. The key point is, given what we know of human nature, it would be unreasonable to expect otherwise. We know that men of God, with monotonous regularity, fall prey to the call of the flesh; why would scientists not worry about the same sorts of things the rest of us worry about – financial reward, respect and admiration, career advancement, buying a new car, travel, fame, money to send the kids to college, and so on?

Thus, it stands to reason that if we consider scientists to be a sample of humanity in general, there is probably just as much lying, cheating, stealing, falsification, obstruction, and downright fraud among scientists as there would be in society in general or in any large subgroup of society.

As a young man in my early twenties, I firmly believed that academics – all academics, but particularly philosophers – were a higher order of human being. Thrust by circumstance into teaching my first college course at twenty-one, I could not wait to attend my first faculty meeting to sit at the feet of the great men who had taught me philosophy, to hear their unguarded reflections on the nature of truth and beauty, perhaps even to add some minor point of my own – thud! Such ruminations were forever laid to rest when I in fact attended my first meeting. The topic of discussion? The fact that the university administration wanted the department to move to another building, where there were *no faculty bathrooms!* Where professors would pee directly adjacent to their own students! Thus, rudely, my naïve hopes were dashed. Many years later they remain dashed. While holding joint appointments in two science departments and a philosophy department, I have yet to attend a meeting devoted to truth, beauty, or other lofty goals. Most often the topics are funding, raises, getting around the dean, or intradepartmental squabbling. In one science department I know, members cannot even hold a civil meeting, and one faction maintains an Internet Web site attacking the other faction!

I am not attacking scientists or philosophers, merely reminding all of us that such people are as human as everyone else – be we ministers, real estate agents, or soccer moms. Professional seekers of knowledge and truth are no exception.

Furthermore, as Wade and Broad point out, science is a *career.* One cannot seek truth without a *job,* or an academic or (increasingly) corporate or government home. And to get a home, let alone a good home, one needs to *produce.*

Consider the average scientist's career path. As a youth, I believed that one chose a Ph.D. topic on the basis of one's interest, completed the dissertation research, acquired a job, and studied whatever drew one's curiosity for the rest of one's life. Reality is quite different. One may indeed be drawn to a science field based on one's interests. But, unless you are independently wealthy, you will need funding to do your doctorate. Such funding usually comes from your adviser's grant money. He or she has little desire to nurture your academic curiosity; an adviser needs students to pursue a smaller topic within the field they are funded to study. While you may wish to study pain in vertebrates, if your adviser is funded to study the genetics of nociception in squid, you will work on some small area of his or her bigger question or you won't be funded. (Even if you pay your own way, few advisers will take a student deflecting them away from their main interests.) Thus your success is inextricably linked with your adviser's goals, funding, and success. When you finish, you will be hired to spin out work in the field you have been trained, though you may have little interest in the area. If you don't move the field forward by getting grants and publishing, you won't last, and you won't have the wherewithal to publish. And so on.

By now you may be married and have a child. You are not only a scientist but a breadwinner and head of a household. Publish or perish is a cliché, but no less true for being so. Your interests have necessarily narrowed to the area you can be successful, lest you are out of the field.

Worse, quantity of results is far more important than quality, because quality is so damn hard to measure, but quantity is easy to weigh. (I once watched a provost defend a tenure decision by holding two candidate files, one in each hand. He proceeded to treat his hands as scales, the heavier file moving down while the lighter moved up, all the while intoning, "How can I promote person B (the lighter file) if he is in competition with A (the heavier file)?") I recall an article in the *Annals of Internal Medicine* commenting on the many cases of people who had stolen other people's work or fudged data and bemoaning the failure to judge people qualitatively by their best work, not the weight of the total.[2]

I am not saying that all scientists therefore cheat or that most scientists cheat. I am saying that strong pressures to be successful provide vectors encouraging cheating or data fudging or dropping the few examples of data belying your hypothesis. When asked off the record in conversations, most scientists will admit feeling these sorts of pressures, and many will relate cases where they are certain colleagues are cheating or fudging.The archetypal case of careerism driving fraud is the case of John Darsee.

John Darsee was something of a wunderkind. By age thirty-three, he had already published over a hundred papers and abstracts done at Emory and at Harvard. At the time the scandal broke, he was a fellow in the lab of noted scientist Eugene Braunwald, who pronounced him the most brilliant fellow he had ever had in his lab.

Three co-workers saw Darsee fake data from an experiment, which they reported to the laboratory director. The director in turn asked Darsee for his raw data, whereupon Darsee faked the raw data. Darsee admitted to falsifying the data, but denied any other fraud. Braunwald, after an investigation, concurred. Though Darsee's NIH fellowship and faculty position were withdrawn, he continued to work at the lab. An NIH committee set up to investigate Darsee, however, produced a radically different result, concluding that nearly every paper Darsee produced was faked. As a later report states, "the committee criticized Harvard's handling of the investigation, as well as the lax supervision

[2] Angell, "Publish or Perish: A Proposal."

and poor record keeping which was under the direction of one of the nation's leading cardiologists."[3] Darsee turned out to have a history of data falsification going back to his undergraduate days at Notre Dame. In a letter to Braunwald, Darsee said, tellingly, "I had put myself on a track that I hoped would allow me to have a wonderful academic job and I knew I had to work hard for it."[4]

Darsee, barred from NIH funding and serving on NIH committees for ten years, went into medical practice; he received little more than a slap on the wrist in the face of having perpetrated one of the paradigm cases of fraud.

Hard data on how often scientists think data falsification occurs among colleagues, let alone how often it does occur, is very scarce, understandably. First of all, it is clearly embarrassing to the scientific community, as well as violative of the ideological belief that scientists are above that sort of thing. Second, as we shall discuss later, scientific ideology believes that the system of science is insulated against damage by cheating. Third, as we shall again see, people who do cheat are in essence seen as "mentally ill." A study that was presented at the American Psychological Association annual meeting in 1987 that reported that one out of the three scientists at a major university suspected a colleague of falsifying data.[5] Wade and Broad recount a 1976 study from the *New Scientist* showing that 92 percent of the journal's readers said that they knew of cases of cheating directly or indirectly.[6] They also cite a 1980 study where chemical engineers were asked what they would do if told by their boss "to invent data and write a report saying that one catalyst was better than another, even though the real data pointed to the opposite conclusion." Only 42 percent said they would not write the report because it was unethical to do so.[7]

Writing in 1991, Phinney speaks of the issue as follows:

> Although it is not possible to achieve the ideals of science with dishonest reporting of research methods or results, it appears to be quite possible to be dishonest in these and many other aspects of one's professional and personal behavior and still have an outstanding career in science. This is evidenced by

[3] Bell, *Impure Science*, p. 112.
[4] Http://www.unmc.edu/ethics/data/darsee/htm, accessed October 8, 2005.
[5] Ft. Collins *Coloradoan*, August 30, 1987, p. A5.
[6] Broad and Wade, *Betrayers of the Truth*, p. 85.
[7] Ibid., p. 86.

the increasing number of highly successful scientists who have been accused of lying, cheating, and or stealing in their scientific work.[8]

There is some evidence that falsification of data begins in undergraduate laboratory, as occurred with Darsee. In a recent paper, Davidson and her colleagues, while teaching a course in "Professional Values in Science," discovered that "over 60% of the students openly admitted to manipulation of data in undergraduate laboratories."[9] This finding led the authors to survey over 700 students in seven undergraduate biology and chemistry courses. They discovered:

Between 40.4 and 75 percent of students in the surveyed course admitted to manipulating data "almost always," and another 20–43.9 percent admitted to such manipulation "often." Students reporting data manipulation "seldom" represented less than 5 percent of those surveyed, and only one student out of over 500 who responded to this survey replied "never." Using correlation analysis, we learned that admission of manipulation in the course surveyed was strongly correlated to admission of manipulation in other courses and our results show a very strong tendency among undergraduate science students to manipulate or make up data when writing laboratory reports. As high as these percentages are, they are similar to results observed in surveys of cheating on tests which Cizek has described as "remarkably and uniformly high." In surveys taken from 1970 to present, from 42 percent to over 90 percent of students reported cheating on tests by self or others (reviewed by Cizek). Out of 6000 students in 31 universities surveyed by Meade, 67 percent of science majors reported cheating on tests. Most surveys of college test cheating ask only whether the student has ever cheated. Our survey expands this question to evaluate how frequently data manipulation in laboratory reports occurs, allowing us to differentiate between occasional events and habitual cheating. Although there are many studies of cheating on college tests, to our knowledge our study is unique in examining manipulation of laboratory data by undergraduates.

Data manipulation apparently does not diminish as the students progress to upper division courses or from non-major to major courses. Commitment to a major subject, presumably because the student intends to continue in this area of science for all or part of his/her professional career, apparently does not diminish this practice.[10]

[8] Phinney, "Toward Some Scientific Objectivity in the Investigation of Scientific Misconduct."

[9] Davidson et al., "Data Manipulation in the Undergraduate Laboratory."

[10] Ibid.

These results buttress our claim that "scientists are people" – data manipulation occurs about as often as cheating does in the general student population. When one adds the requirements of science as a career discussed earlier, the presence of cheating in science should not surprise us, though it most certainly should upset us and tarnish science's view of itself as impervious to duplicity because of its commitment to seeking the truth.

In addition to careerism, there are a number of other vectors conducive to scientific misconduct in the broadest sense. One of these is money. In today's scientific milieu, discoveries may be worth a fortune, and increasing numbers of scientists create start-up companies funded by venture capitalists based on their results.

Earlier in our discussion of human research we mentioned the death of Jesse Gelsinger, the youth who participated in a gene therapy trial at the University of Pennsylvania. A detailed *New York Times* article[11] mentions that though Gelsinger is officially the first person to die through the failure of gene therapy, other unreported deaths have occurred. Large amounts of money hinge on the success of gene therapy, and thus damning data is withheld both from patients (as mentioned earlier) and from the public.

Harold Varmus, NIH director, "had made no secret of his distaste for the conduct of gene-therapy researchers. He thought the science was too shoddy to push forward with human testing.... It irked him to have to sign off on protocols the [National Recombinant DNA Advisory Committee (RAC)] approved, and it irked him even more to see biotech companies touting those approvals, like some kind of NIH imprimatur, in the business pages of the papers. 'Somedays,' said the committee's former executive director,... 'it felt as though the RAC was helping the biotech industry raise money. Dr. Varmus hated that.'"[12] Nonetheless under pressure from the industry, the RAC was neutered. Even though some members had doubts, the Gelsinger experiment proceeded with the tragic results we have recounted.

After this tragedy, in response to a query from NIH, "gene therapy researchers all over the country revealed more than 650 dangerous adverse reactions they had previously kept secret, including several

[11] Stolberg, "The Biotech Death of Jesse Gelsinger."
[12] Ibid.

deaths. . . . Industry representatives are bridling at proposed rules that would prevent them from wrapping adverse reaction reports in pages of trade secrets as they do now, thus keeping them confidential."[13]

Paul Gelsinger, Jesse's father, is quoted as saying, "This is all about money, prestige."[14] It is impossible not to believe that financial concerns did not contribute to Jesse's death. In my own research for this book, I talked to a person who had worked in the University of Pennsylvania lab where Jesse died. The person told me, "I left the lab before the tragedy... because I was sure someone would be killed."

In the same vein, there are many cases of drug company scientists concealing data that show dangers associated with new drugs. A very recent case involved whistleblower Dr. David Franklin of Harvard Medical School, who, while working for Warner-Lambert, found himself in the thick of promoting a drug called Neurontin.

> It had only been officially approved for the treatment of epilepsy, but his [Franklin's] job was to persuade doctors of its suitability for at least a dozen other disorders. Company tactics included paying doctors to appear as authors of journal articles on the wider use of Neurontin, that had actually been written by employees of the drug company, as well as paying for hundreds of doctors to attend expensive dinners and weekend retreats where they were urged to prescribe Neurontin. When Franklin found evidence of side-effects in some children he was ordered not to tell doctors about it. At the end of the year, non-epilepsy conditions made up 78% Neurontin's $1.3bn sales.[15]

The article goes on to cite another disturbing fact:

> An alarming study published in the *American Medical Journal* in February looked at the financial interests of the authors of clinical guidelines – authoritative reports put out to guide doctors on the best treatments on common conditions such as asthma, heart disease and ulcers. The study found that, on average, each of the authors had links with 10 companies, including the ones whose products they were recommending. Only one of 44 carried a declaration of the authors' competing conflict of interests, the report concluded, because there were too few independent guidelines to make a comparison.[16]

[13] "Bioethics: Gene Therapy Business: The Tragic Case of Jesse Gelsinger," *News Weekly*, August 12, 2000.
[14] Ibid.
[15] Burne, "The Drug Pushers."
[16] Ibid.

Another article, reflecting on the Neurontin story, points out:

> Drugs for abnormal heart rhythm introduced in the late 1970s were killing more Americans every year by 1990 than the Vietnam War. Yet early evidence suggesting the drugs were lethal, which might have saved thousands of lives, went unpublished. Expensive cancer drugs introduced in the past 10 years and claiming to offer major benefits have increasingly been questioned. Evidence published in the *Journal of the American Medical Association* showed that 38% of independent studies of the drugs reached unfavourable conclusions about them, compared with 5 percent of the studies paid for by the pharmaceutical industry.[17]

In one of the best books on scientific corruption, *Impure Science: Fraud, Compromise, and Political Influence on Scientific Research* (1992), author Robert Bell has a long chapter on abuses stemming from the pharmaceutical industry. I used this book very successfully to teach a science and ethics course, but it is unfortunately out of print. In that chapter Bell focuses on Zomax, a painkiller capable of generating anaphylaxis; a heart valve device susceptible to fracture; and Versed, a sedative causing death via respiratory depression. In all three cases, relevant evidence was withheld by the pharmaceutical companies.

With biomedical research being funded ever-increasingly by private companies – even university research with large amounts of money at stake – clearly strong incentives exist to conceal recalcitrant data. This, of course, encourages secrecy as well as research oriented to please corporate sponsors.

Thus, money joins with careerism to create a fertile climate for misconduct, concealment, and falsification. Yet another ingredient in this unsavory brew is entrepreneurship. Major researchers often have numerous grants, and it is impossible for them personally to supervise all the work done. Thus, they hire large numbers of personnel and are themselves rarely present in a given lab. (As Braunwald did, for example, they hire senior scientists as lab directors.) Yet the ultimate responsibility lies with the principle investigator. Thus, a senior researcher with his or her name on a publication may in fact know little of how the experiment was conducted but bears responsibility for fraud (or misconduct) if it is exposed. Such an event occurred in 1980 when Philip Felig, a leading entrepreneurial researcher at

[17] Laurance, "Licensed to Kill – Facts the Drug Firms Conceal."

Yale, coauthored a paper with Vijay Soman, who had written twelve papers for which data were missing or fraudulent, and all of which were eventually retracted. It turned out that the paper coauthored with Soman was in fact plagiarized from a paper that Felig had been asked to review but passed on to Soman. Though Felig himself had falsified nothing, a job offer to serve as chairman of medicine at Columbia was withdrawn.[18]

I happened by sheer chance to be in the dean of medicine's office at Columbia when the story broke in the *New York Times.* He told me that though Felig was largely guilty only of lack of oversight, his career was irrevocably blemished.

A similar case occurred in 1996 in Britain in the field of obstetrics. Malcolm Pearce, a world-famous expert in obstetrics, published a paper in the *British Journal of Obstetrics and Gynecology*, of which he was an assistant editor. The paper covered a case of allegedly reimplanting an ectopic pregnancy, resulting in a baby being born. Pearce listed Geoffrey Chamberlain, editor of the journal, president of the Royal College of Obstetricians and Gynecologists, and head of his department at St. George's hospital, as a second author. A second paper by Pearce also appeared in that issue. A *Daily Mail* story exposed the two papers as fraudulent, and presented Chamberlain as saying he had not really done work on the study but had received "gift authorship." Among other things, the alleged patient did not exist, nor did the patients used in the second paper. Pearce was fired, but Chamberlain had to resign all his positions, "a terrible end to a distinguished career," as an author who described the case put it.[19]

It is worth pausing to examine the concept of gift authorship. If one looks at a group of typical science research papers, one will invariably find a long list of co-authors. Other than the first author, one cannot reliably judge the contribution of any other listed authors. Someone may have supplied a reagent and gotten their name on an author list; someone else an instrument; on occasion, people do not even know they have been listed! I myself have appeared as an author on papers I have never seen; if you held a gun to my head I could not tell you what they are about.

[18] Smith, "Research Misconduct and Biomedical Journals."
[19] Nolan, "Guidelines on Good Research Practice."

This relates to scientific misconduct in three ways. In the first place, it cheapens publication if one can appear as the author of a paper one doesn't know about. This, in turn, can (and does) encourage cynicism about publication value. (My science colleagues are amazed when I tell them that out of over 300 publications of mine, only half a dozen have other authors, usually only one.) In the second place, it encourages fraud or rather seems to take it for granted. After all, if I can take partial credit for loaning you a chemical, why not fix this little piece of recalcitrant data or number? Third, when a paper is revealed as having resulted from misconduct, everyone whose name is on it, for whatever small reason, is stigmatized to some extent.

Yet another source of data falsification comes from paradigm blindness, wherein one is so certain of the truth of a belief, usually a theory, that data don't matter. There is a long history in science of ignoring recalcitrant data if a theory seems otherwise flawless. The ether as a medium for propagation of light waves in physics was so well established in the nineteenth century that when the Michelson-Morley experiment showed that the speed of light both with and against the "aether wind" was exactly the same, no one worried about it, dismissing the results as error in measurement that would be corrected sometime in the future by more precise instruments.

Thomas Kuhn[20] and Paul Feyerabend[21] have pointed out how easy it is to turn recalcitrant data putatively falsifying a theory that otherwise works over a broad domain of issues into a "research problem." Thus while Newtonian physics could explain a great many phenomena, it could not account for behavior of the moon. To committed Newtonians, this became a problem for future research, not a refutation of Newton.

Another example comes from Sir Arthur Eddington's attempt to verify general relativity. He traveled to an island of Africa to watch the solar eclipse of May 1919. Relativity predicted that light would bend near a massive star as a result of gravity, and thus stars would appear shifted away. Eddington confirmed this by observation. When reporters asked Einstein what he would have said had Eddington's data

[20] Kuhn, *Structure of Scientific Revolutions.*

[21] Feyerabend, *Against Method.*

not turned out to support relativity, he replied in essence, "So much the worse for the data – the theory is correct!"[22]

A final example comes from the work of Gregor Mendel on heredity in pea plants. Sir Ronald Fischer, the father of statistics, showed that Mendel's results were determined by his theoretical commitment. When Mendel looked at the F_2 generation of peas he had crossed, he found that the resulting phenotypes fit his predicted ratio of 3:1. Fisher found that if he did the experimental crosses 100 times, Mendel would have only an under 5 percent chance of achieving the 3:1 ratio. As one author puts it,

> Plausibly, Mendel kept counting seeds until the ratios were convincingly close to the prediction, then he stopped. (At this time, this was thought to be perfectly good experimental procedure – even now it must be very tempting to stop an experiment as soon as the data look good enough to persuade the researcher that he or she was right all along.)[23]

In sum, a powerful vector for fudging data or smoothing data is theoretical bias. If one is firmly committed to a theory that seems to explain a given phenomenon very well except for a few "minor aberrations," one is sorely tempted to eliminate the aberrations or ignore them until such time as "better" data can be found. We cited a very elementary example of this when we discussed seeing with one's expectations earlier in this book.

To borrow an idea from Leon Kamin, imagine a scientist who is testing a hypothesis on 1,000 rats. Some 990 rats show exactly what was predicted, ten rats do not. There is surely a great temptation to disregard the ten as ill or somehow anomalous, particularly if one needs the publication, strongly believes in the theory in question, or both.[24]

Some time ago I saw this phenomenon in action. Our veterinary school was interested in the question of whether it was necessary to learn surgery on a live animal or whether cadavers would serve just as well. One resident in surgery had done some ingenious research

[22] Clark, *Einstein: The Life and Times*, p. 287.

[23] "Gregor Mendel and the Discovery of Genetics," http://www.unbf.ca/psychology/likely/evolution/mendel.htm, accessed October 8, 2005.

[24] Kirtz, "Healthy Skepticism."

on this, where he taught one group of surgically naïve students the standard surgery rotation on live animals and another group using cadavers. He then had both groups do an untaught surgical procedure, videotaped them, and then, in a double-blind procedure, sent all the tapes to board-certified surgeons asking them whether they could tell which students had trained on live animals and which on cadavers. The result? No difference in perceived skill.

The researcher called me and asked me to come to his seminar. When I asked why, he said, "I need your protection! The surgeons won't like my results." I went, and sure enough, one very prestigious surgeon immediately began to attack the researcher's methodology very aggressively. I interrupted him by saying, "Wait a minute! Would you be attacking his methodology if the results had come out the other way, in favor of your bias?" To his great credit, he stopped dead. "You're so right!" he said. "I'm supposed to be a scientist and follow the data, not my biases."

A final reason for data falsification can possibly be found in one of the most basic of human traits: laziness. In the 1970s, the federal government through the FDA was contracting a great deal of toxicology out to private laboratories. In an audit of 867 laboratories, bad data and bad study conduct were found in 618: Lab books were sloppily kept, sick animals not separated from healthy. Four laboratory managers were found guilty of fraud. As a result, in 1976, the government passed the Good Laboratory Practices Act, *which essentially made into federal law what ought to be presuppositional to doing science*: better record keeping, health checks, raw data preservation, and so on.[25] One of my friends, a high official at CDC, had called this law "the shame of the scientific community," that it was shameful that one needs to legislate what should be automatic to doing science. Yet it turned out to be the case. I would therefore conclude that at least some data mismanagement in science is the result of sloppiness and/or laziness, not the more complicated reasons we have discussed earlier.

In any case, we have detailed some of the very complex factors that can result in misconduct in science in the broadest sense. How does the scientific community respond to revelations of misconduct? In their 1982 book, Wade and Broad exemplify what even then were standard

[25] Cook, "Good Laboratory Practice (GLP) versus CLIA."

responses. They quote Lewis Thomas as saying that cases of fraud "can be viewed as anomalies, the work of researchers with unhinged minds." Philip Handler, head of the National Academy of Sciences, told a 1981 congressional committee investigating misconduct that "one can only judge the rare such acts that have come to light as psychopathic behavior originating in minds that have made very bad judgments – ethics aside [NB!] – minds which in at least this one regard may be considered deranged."[26]

The editor of *Science*, Daniel Koshland, has affirmed in 1987 that "99.9999% of [scientific] reports are accurate and truthful,"[27] the implication being that fraud amounts to "spitting in the ocean," that is, doesn't make much difference. This was echoed in a paper by Sharoni Shafir and Donald Kennedy (former FDA commissioner, president of Stanford, and editor of *Science*), wherein the authors attack the media view that misconduct is rife.[28] But such claims miss some very important conceptual points.

To understand the first point, let us recall Descartes'discussion in Meditation One regarding the validity of sense experience. The argument *seems* to be that since the senses sometimes deceive us, we can never trust them. But his claim is far more subtle. What he is in fact claiming is that we use the senses to correct the senses and that false sense experiences are not qualitatively different from true ones; that is, don't wear a "badge" saying "false sense experience." (If they did, they couldn't fool us.) So Descartes' point is there is no infallible, non-circular way to detect truth and falsity in sense experience. Exactly the same holds of tainted experiments. It is not the number of them that is the point; it is rather that other scientific work builds on them, so if we can't know for sure which reports are tainted, we can never know how many other reports, building from the tainted ones as a base, are also tainted.

A related point grows out of Pierre Duhem's famous argument that there can be no crucial experiments in physics.[29] A crucial experiment C is meant to decide definitively between two competing hypotheses,

[26] Broad and Wade, *Betrayers of the Truth*, p. 60.
[27] Quoted in Phinney, "Toward Some Scientific Objectivity."
[28] Shafir and Kennedy, "Research Misconduct: Media Exaggerates Results of a Survey."
[29] Duhem, *The Aim and Structure of Physical Theory*.

A and B. Suppose the result favors B, and seems to disqualify A. Since A rests on a variety of other presuppositions, we cannot definitively say that C disconfirms A, rather than some other presupposition on which A rests. In other words, because any experiment testing a hypothesis presupposes a whole range of theoretical assumptions, it may well be that the experiment is exposing a flaw in one of these, rather than in the hypothesis being tested. Similarly, if we don't know what data current experiments rest on, and we are testing a hypothesis founded on that data, and some of that data may be false, we cannot know whether the experiment simply disqualifies the hypothesis being tested or points up the presence of flawed data in the tissue of historical presuppositions to any experiment.

If these two points are correct, a very small number of falsified experiments can wreak major havoc with subsequent science. The 1993 Commission on Research Integrity reported fraudulent research on "breast cancer, immunology, neurological disorders, Alzheimer's, fetal drug addiction, anorexia, arthritis, and toxicity."[30] In other words, fraudulent research occurs in areas that may have a major negative impact on human health.

Finally, even a small number of well-publicized cases of fraud can irrevocably damage public support for science, which is at the moment not all that strong or stable. In addition, it can discourage bright young people from choosing research as a career. (I have been told by friends at NIH that for some time bright young women in high school have indicated that they do not wish to pursue a career in biomedical research because they don't wish to "hurt animals.")

The "official" scientific community response to the increasing number of misconduct allegations, as we mentioned, is to deny their significance. In addition, there is a strong tendency to cite institutional self-policing mechanisms built into science that assure adequate management of misconduct. With some justice, scientists fear external meddling in science by people who are not scientists and don't understand how science operates. The current social craze for "accountability" has certainly done more harm than good in universities, as administrators proliferate and demand endless reports (to justify their existence),

[30] Carson, "Striking a Blow for Scientific Integrity."

which in turn creates volumes of useless work for faculty that cut into time for research and teaching. When I began to teach, one told one's department head what one did in scholarship, teaching, and service. Now one fills out endless forms showing what one had done for "outreach," "diversity," "service learning," and all the other hot-button buzzwords. So it is not irrational to resist external intrusion into science. But the price of autonomy is responsibility. So, it is fair to ask, what are science's self-policing mechanisms, and do they work adequately? These mechanisms are generally taken to be peer review for research proposals, the referee system for reviewing manuscripts, and the in-principle replicability of all scientific results.

How well do these work? Let us begin with peer review. Currently, the federal government funds a major part of scientific research. Obviously, there are always more proposals than there is money. Proposals are therefore subjected to review by panels of experts or "study sections" that score the proposals in terms of scientific merit. Obviously, peer review is prospective (though many researchers apply for grants after they have some indication that their idea works). The obvious question that arises is, simply, how can a prospective procedure assure that the actual execution of the work will not involve misconduct or data fudging or falsification? What it should be able to catch is plagiarism. But peer review creates its own plagiarism problems, most notably, the problem of reviewers stealing ideas or approaches from the proposals they review. Since review is anonymous, no one can prove that his or her ideas were stolen by someone else.

Another problem in peer review is that it is inherently very conservative in what is funded. A field is made up of experts whose work defines the field. For example, one of my friends is a well-known physiologist. He once told me that, in submitting grant proposals, he had been turned down only once in his career. This occurred in a case where if what he proposed to study turned out to be confirmed, some major dogmas in the field entertained by the experts, that is, the reviewers, would turn out to be falsified. Needless to say, human beings (even scientists) typically do not dig their own graves, let alone cut their own throats; so he was not funded.

One of my own colleagues experienced this sort of conservatism with a vengeance. He believed that in planaria he had found a nonmammalian model for toxicology of heavy metals, for planaria behave erratically in the presence of heavy metals. If he was correct, thousands of assays could be done far more cheaply than by the use of rodents; in fact, thousands could be done on top of his desk. Furthermore, much pain and suffering would be spared. When he submitted his proposal to the toxicology review boards, he was told that toxicology was done on rodents and to try psychology. When he submitted there, he was told that the group has no interest in toxicology. He never did receive funding for this promising idea.

Given my interest in alternatives to animals in research, peer review is particularly vexacious, as "experts" in a given field develop their own ways of doing things. If the "standard way" is live, animal-based, one finds formidable barriers to replacement by nonanimal means. An excellent article on this issue by Andrew Rowan may be found in my book on animal use in biomedicine.[31]

The second institutional vehicle for self-policing identified by the scientific community is the referee process. Research, of course, is of no value to the researcher, science, or society unless published. All journals utilize experts to review submitted papers. These referees are supposed to catch misconduct. While one can certainly imagine referees catching cases of blatant plagiarism, they, like grant reviewers, have no ability whatever to assure proper conduct of the research. All they can do is assure that design material and methods and procedures are appropriate. Even in terms of catching plagiarism, that too is unlikely with thousands of medical journals across the world. If a cheater chooses carefully from obscure and foreign journals, he or she can probably publish these pieces as his or her own with little chance of discovery.

Many of the same problems we have recounted earlier regarding grant reviewers beset the referee system. For one thing, I have heard very many stories from scientists claiming that their ideas were appropriated by reviewers who in fact rejected their papers. Since the review process is anonymous, such claims can never be verified.

Again, the issue of reviewer bias obtrudes as it does in grant proposals. In fact, grant reviewers and journal referees may be the same people! They are, after all, the experts. In my experience (which

[31] Rowan, "Alternatives to Animals: Philosophy, History, and Practice."

admittedly is largely with philosophy articles, though I have published science papers as well), manuscript review tends to be totally capricious, whimsical, and arbitrary. Two reviewers of the same manuscript may say dramatically opposite things.

One of my colleagues sent a scientific manuscript to a scientific journal, and received a rejection. Among the reviewer's comments were "this paper presupposes evolution and not all of us believe in evolution!" Other colleagues who wrote on pain found reviewers criticizing the assertions made that animals felt pain. I have had manuscripts of mine containing philosophical arguments rejected by scientific journals because I had insufficient *data*!

One of my favorite experiences concerns a paper I wrote as a very young professor, addressing a neglected issue in the history of philosophy. I submitted the paper to what is unquestionably the best journal in the field. The paper was rejected for a set of reasons. I then proceeded to submit the paper to ten other journals, all of whom rejected it, but for logically incompatible reasons. I never received the same criticism twice. This process occurred over a period of years. By then, the editor of the "best journal" had died and been replaced, so I resubmitted just for fun. The paper was quickly accepted, and became a small classic in the field!

Wade and Broad cite some other studies devastating to the integrity of the review process.

> A more elaborate investigation was designed to scrutinize referees' theoretical biases in assessing papers. Michael J. Mahoney had a journal send out fictitious manuscripts on a hotly debated aspect of child psychology to seventy-five referees whose personal positions on the problem were on record. The manuscripts all described the same experimental procedures but the purported results were different, some favored the reviewer's perspective, some refuting it. The result: "Identical manuscripts suffered very different fates depending on the direction of their data. When they were positive (i.e., in accord with the referee's particular bias), the usual recommendation was to accept with moderate revisions. Negative results earned a significantly lower evaluation. . . ."

By accident, a glaring error was introduced in the manuscripts sent out for review. The error was not spotted equally by all reviewers: In manuscripts with "positive" results, only 25 percent of the reviewers noted it; but the error was obvious to 71 percent of the reviewers when the results were "negative," that is, inconsistent with the reviewer's theoretical perspective.[32]

A wonderfully witty poem by Ralph Boas summarizes some of the concerns about reviewers:

The Author and the Editor

I send the paper in
They say it is too thin.
When I've corrected that
They say it is too fat.
The next word that I hear
"It isn't really clear."
But after I explain
They look at it again.
At last they write to say
There's been so much delay
That with regret they find
(And hope I will not mind)–
Admittedly, it's sad –
But someone else just published everything I had![33]

The power of journal editors is illustrated by the following anecdote: Some years ago in the early 1980s, I was invited to give the banquet speech to the annual meeting of the Shock Society, an international group of researchers who study circulatory shock. Knowing that I was an animal ethics person, young researchers approached me with a concern. To publish animal research on traumatic shock in the chief journal in the field, *Shock*, one was required to use unanesthetized animals, because they were modeling unanesthetized humans who experience trauma. The researchers felt that this was animal abuse, and had received harsh criticisms from staff at the conference center who had seen films of such experiments during the course of the conference. I raised the issue in my speech and showed the audience many other parameters that could skew shock research, for example, how the animals were treated, yet were not taken into account in journal policy. I challenged them to show that one could not map the responses of anesthetized animals onto unanesthetized animals and that failure to

[32] Broad and Wade, *Betrayers of the Truth*, p. 103.
[33] Boas, "The Author and the Editor."

control pain was ethically unacceptable. When I finished, the journal editor stood up and said, "I'm glad you brought that up! I was just about to announce that we will henceforth require anesthesia for trauma."

The last official safeguard is "replicability." The argument is that any scientific result should be able to be replicated by others and stand or fall with the results. This is particularly relevant to research suspected of involving misconduct. There is something to this claim: A clear example occurred in the infamous cold fusion scandal. In March 1989, two researchers, Martin Fleischmann and Stanley Pons, announced that they had accomplished "cold fusion," paving the way for limitless power at little cost. On the basis of this experiment, the researchers requested $25 million from the federal government to continue this research. Shortly after, however, reports began to flood in from laboratories around the world who had attempted to replicate these results. Some who claimed to have achieved replication recanted and admitted to faulty equipment or measurements. The consensus is that there were not reproducible cold fusion results, and mainstream science has dropped the issue, even though commercial interests continued for a while to support Fleischmann and Pons.

Interestingly enough, there are still people who believe that cold fusion was deliberately discredited by hot-fusion types or buried by the same scientific community that buries psychic powers and UFOs.[34]

Whatever one's view of cold fusion, it provides a paradigm case of a situation where replicability is necessary. Here was a paradigm-defying result with enormous commercial potential. In such a case, money becomes available for replication. But suppose the experiment in question is a piece of pedestrian science on which little or nothing hangs? There would be no great rush to replicate it and no money available to do so. And, in essence, no one would care. If the researcher is a clever cheater with nothing major at stake, no one will try to check his or her results – no one wins fame or tenure for checking trivial but plausible results. (Most papers are never cited after publication.)[35] At the same time, if one does try to replicate a result and fails, that is no proof that the work was fraudulent – there are innumerable other variables that could have been neglected or passed over.

[34] Voss, "Whatever Happened to Cold Fusion?"
[35] Sample, "Giants of Science."

Wade and Broad go so far as to affirm that "[t]he notion of replication, in the sense of repeating an experiment to test its validity, is a myth, a theoretical construct dreamed up by the philosophers and sociologists of science."[36] In other words, replication is yet another component of scientific ideology.

What is in fact probably the best way to detect misconduct is via the work of whistleblowers, people who witness such behavior and report it to someone who must act. Whistleblowing in a public way assures that the establishment will not bury a concern. Walter Stewart, who, along with his colleague, Ned Feder, was a thorn in NIH's side by maintaining a web site encouraging whistleblowing, has been a strong advocate of such activity as "tak[ing] the responsibility for the self-regulation of science. If we as a community do not, someone else will."[37] Ironically, Stewart and Feder, both of whom worked for the NIH, had their laboratory shut down in May 1993, their documents seized, and were reassigned to areas outside their training. Such, as we shall see, is too often the fate of whistleblowers.

The first thing to realize about whistleblowing is that our American social thinking is very ambivalent about it. On the one hand, Daniel Ellsberg was seen as a hero when he blew the whistle on the Pentagon papers, and Erin Brockovich became a folk hero and subject of a movie for blowing the whistle on corporate pollution and leading a class-action lawsuit. On the other hand, people are not generally comfortable with "squealers," "rats," "stoolies," or "tattle-tales." I grew up in Brooklyn in the 1950s, and if we had "a beef" with another student, we were never supposed to tell the teachers. We handled it ourselves. My son grew up in the "kinder, gentler" 1980s, and came home from third grade one day to tell me proudly that a classmate had hit him and that he told the teacher. It was all I could do not to grab his collar and call him "a damn rat!" In our veterinary school, as in many other institutions, we have an honor code, that requires not only that you not cheat but that you report suspicions of cheating. This almost never occurs and is a serious threat to honor codes.

Not only do people have ethical ambivalence about whistleblowing, they also have justified self-preservational concerns. One of my science

[36] Broad and Wade, *Betrayers of the Truth,* p. 77.
[37] Walter Stewart's Web site.

colleagues recently reviewed a grant and found that it had largely been plagiarized from someone else's published work. She called the relevant authorities and was amazed to find that instead of encountering gratitude, she was treated as if she were the miscreant. To her knowledge nothing was ever done in this case. This is the rule, not the exception. Bell, writing in *Impure Science,* categorically asserts that "[i]n every case examined in this book, the whistle-blower has suffered at least as badly and almost always far worse than the scientist who committed fraud."[38]

Why this should be follows from what we know of human nature. A whistleblower almost by definition needs to be a colleague or coworker. If the whistleblower is successful, the flow of money to the research institution is truncated, projects are aborted, and jobs disappear. Thus the whistleblower is seen as a "traitor" harming his or her peers and institution. Second, people become paranoid about the presence of a colleague whistleblower. Everyone has some skeleton in his or her closet, and the idea that a colleague is capable of exposing yours makes you uncomfortable around such a person. This in turn is impossible to conceal.

Recriminations against the whistleblower may involve firing, shifting his or her job (as we saw in the case of Stewart and Feder), supervisors quietly committing to giving them bad evaluations, promotions stymied, equivocal letters of reference, or other forms of overt retaliation. The Ryan Commission Report on Integrity and Misconduct in Research of 1995 recognizes these problems and articulates a "Whistleblower's Bill of Rights," which includes "a duty not to tolerate or engage in retaliation against good faith whistleblowers,"[39] institutional due process, institutional heedfulness to complaints, and timely investigations. Even if these well-meaning principles were adopted by all research institutions, however, it is difficult to see how one could prove even what we have called overt retaliation, which can always be cloaked in something other than whistleblowing, for example, slowness in completing tasks or failure to get along with colleagues.

Indeed, more so than overt, documented retaliation, the issue is an attitude that will likely be adopted toward the whistleblower that

[38] Bell, *Impure Science,* p. 258.
[39] Ryan, "Ryan Commission Report."

is difficult to document, for example, exclusion of that person from social gatherings, small talk, and general camaraderie. The prison system eloquently attests to how a group can make a person's life miserable without creating an actionable trial; indeed, we can see the same sort of thing happening in junior high schools.

On the other side, a whistleblower may have extraordinary power to harm a person irrevocably with little tangible evidence. Consider the following point: Suppose that every day for a year I march around the quad at my university carrying a large, professionally produced sign that says, "Dr. Jones is not a child molester." Ten years later, when Jones is mentioned, many people will recall vaguely, "Oh Jones! Wasn't he linked with child molestation?" People will feel uneasy around Jones, and his life will be changed for the worse in very tangible ways.

By and large, acts to protect whistleblowers will not do very much – there are extant acts covering the federal government that have not done much for some time. People who are moved by conscience will act; others will not. And we cannot expect that any more than a relatively small group will embrace whistleblowing. Subtract from that group the small number who blow whistles for mischievous reasons, and what remains is at best only a part of a solution to the misconduct in science issue.

The Ryan Commission report makes an excellent point when it affirms that

> the best protection for whistleblowers and witnesses lies not in federal regulation, but in an institutional culture that is committed to integrity in research. To establish such a culture, committed institutional leadership is an essential component. Institutional commitment expedites good-faith resolution of disputes over alleged research misconduct and helps to prevent prolonged adversarial proceedings.[40]

In many ways, my own career provides an instantiation of these sage words. As mentioned in our chapter on animal research, I was hired by the Colorado State University veterinary school to develop the world's first course in veterinary ethics. I have already recounted how this led me to engage practical issues, such as multiple surgery and invasive laboratory exercises. What I have not yet discussed is how

[40] Ibid.

that job evolved into something more than that, a job as ombudsman for animals and indeed ombudsman for a whole range of bioethical issues, and as one-on-one educator for faculty on ethical matters.

This all occurred in two related paths. In the first place, the kind of ethical discussion I was hired to conduct with veterinary students inevitably spilled over into discussion with faculty, as illustrated above on the issues of surgery. Though guarded initially, most faculty were quite willing to meet and discuss the issues in question. I was fortunate to be at a western veterinary school with a significant agricultural influence, permeated with the western ethic of "first you hear 'em out and then you hang 'em," rather than "just hang 'em." I was constantly met with "If you are going to discuss____" (where____ might be surgery, agriculture, or pain control), "at least know what you are talking about. Read this and that." The result was that we learned from each other in very natural contexts.

Second, about the same time I began with the veterinary school (1976), I came to know the university laboratory animal veterinarian, hired to assure proper care and use but with no authority (there were as yet no laws.) As we discussed earlier, we began to work together on writing laws, but also on problems on campus, where I spoke as a colleague to researchers he was having trouble with. Again, they were very open and in many cases would begin to think for the first time about their moral obligations toward research animals.

By the mid-1980s, I had been authorized to speak for the university on animal research and other bioethics matters, such as human research. I had publicized the changes we made in surgery, laboratory exercises, voluntary adoption of an Institutional Animal Care and Use Committee, and so on, and we received much public support for trying to do the right thing. The researchers in turn appreciated the fact that we had public support and that they were not targeted, as occurred in other institutions. The net result was that by 1982, we were written up in *Nature* as "the best in the U.S" on laboratory animal welfare.

Eventually, the vast majority of faculty were ethically aware and felt that ethics was a protection for them, not something irrelevant to science. My position evolved into being an ombudsman on ethical matters, resolving issues raised by students, faculty, and staff in an off-the-record, dialectical, nonadversarial way. To take one example: A former veterinary student of mine doing a Ph.D. came to my office and said he worked in a lab where students without proper training were told to do surgery. This was at 10 A.M. By 2 P.M., after a discussion with the dean and a major department head, we had established a forty-hour course in surgery that any graduate student had to take and pass before they could touch an animal. Everyone thought it was a good idea, particularly because the university paid for it!

In any event, I would strongly argue for an ethics ombudsman on every campus, who is seen as a colleague. As not part of a formal administrative structure, such a person is nonthreatening and can operate in a collegial way, without any cudgel other than reason and peer pressure. This makes ethics a living and dialectical activity, rather than a bureaucratic one. Colorado State University feels so strongly about this role that they have made plans to keep it going after I retire or expire! I can strongly attest to the fact that it works, particularly in the animal use area. (Since we have no medical school, I have not had to deal with life-and-death human research issues.)

Such a position can help to create the kind of institutional culture mentioned in the Ryan report. What we did learn the hard way is that, if not nurtured, such a culture can gradually wither away as senior faculty leave, and nothing is done to transmit that culture to young faculty members. Thus the ombudsman must work to keep the tradition alive and disseminate it. Most satisfying is the fact that though I have had many battles with faculty over ethical matters, none left scars or grudges; most of my adversaries have eventually thanked me for helping to reshape their thinking.

Thus I see the ombudsman position as that of an educator. For years, I harangued the research community with the need for ethical discussion and education in science. After all, I repeatedly argued, one cannot say, "Science has nothing to do with ethics" or "Science is value-free" and then upbraid students (or colleagues) for acting unethically. If science is independent of ethics, why not cheat, falsify data, plagiarize, run trials until they come out as you wish them to, fudge, and so on? Gratifyingly, the NIH has grasped this point and now mandates courses in science and ethics for graduate students on training grants. In my view, only education can solve the problems we have raised in this book, for without education in ethics these issues are not even acknowledged, let alone properly addressed.

Let us return to an early theme orchestrated in this chapter: Scientists are human. And humans, in turn, are subject to a vast array of pressures and temptations. Learning ethics (both $Ethics_1$ and $Ethics_2$) is a historically (though not infallible) proven way of helping to resist these temptations or at least of helping to understand the implications of succumbing to them. In particular, even a little bit of $Ethics_2$ discussion would have raised this question in people's minds: How can you teach students that science is value-free and ethics-free and then be shocked and upset that they falsify, fudge, or plagiarize data, and then claim that such behavior is "unethical"?

Indeed, a basic reading of Plato's *Republic* could have taught scientists the absurdity of denying the relevance of ethics to science. Plato argues that even a band of thieves united together for some nefarious purpose, such as robbing a bank, must have rules and mutual expectations understood by all and binding on all. How could you rob a bank if all parties did not agree to do their jobs, to respect each other's roles, not to lie regularly to each other, not to kill each other, and so on? As ordinary language puts it, there must be "honor among thieves." In other words, any group of people united for a common purpose must have some rules binding on all parties – some "thou shalts" and "thou shalt nots," in order to regulate their conduct – that is, must have some semblance of an *ethic.*

So once again, as I did in the chapter on human research, I would argue the need for education of scientists in ethics. This can be done in many ways, ranging from full-blown courses covering the sorts of matters we did in this book to informal discussions among members of research groups. The key points of such education are to examine scientific ideology and to make ethical thinking second nature to young scientists.

I know from both my own thinking and my own teaching that stimulating thought on ethical matters in science is not difficult once one circumvents ideology. In my own case, it was far harder for me to learn to embrace scientific ideology, since it was so counterintuitive, than for me to identify ethical issues in science. Having taught science and ethics in various forms for almost thirty years, I know that today's students take to it very naturally, having grown up in a society wherein many ethical questions about science have been raised even in secondary school. Virtually every young scientist will have been exposed to questions about racial and gender bias in research (*vide* the Tuskegee syphilis studies and the emphasis on males and male diseases in biomedical research), questions about dissection and harm to animals in research, and probably to questions about scientific misconduct.

The failure to teach young scientists to think and reflect about ethics is an intellectual and prudential sin, one punishable by loss of scientific credibility in society, loss of integrity in the very processes presuppositional to doing science, lack of social appreciation for scientists, and by a diminution of science's potential as a force for good in society.

Bibliography

Anand, K. J. S., and K. D. Craig. "New Perspectives on the Definition of Pain." *Pain* 67 (1996): 3–6.

Anand, K. J. S., and P. R. Hickey. "Halothane-Morphine Compared with High Dose Sufentanil for Anesthesia and Post-Operative Analgesia in Neonatal Cardiac Surgery." *New England Journal of Medicine* 326, no. 1 (January 2, 1992): 1–9.

Anand, K. J. S., and P. R. Hickey. "Pain and Its Effects in the Human Neonate and Fetus." *New England Journal of Medicine* 317, no. 21 (November 19, 1987): 1321–29.

Angell, Marcia. "Publish or Perish: A Proposal." *Annals of Internal Medicine* 104 (1986): 261–62.

Aristotle. *Metaphysics.* In R. McKeon (ed.), *The Basic Works of Aristotle.* New York: Random House, 1941.

Ayer, Alfred Jules. *Language, Truth, and Logic.* London: Victor Gollancz, 1946.

Beecher, Henry K. "Ethics and Clinical Research." *New England Journal of Medicine* 274, no. 2 (1996): 354–60.

Beggs, S. "Postnatal Development of Pain Pathways and Consequences of Early Injury." *Proceedings of Eighth World Congress Veterinary Anesthesia 2003.* Knoxville: University of Tennessee, 2003.

Bell, Robert. *Impure Science: Fraud, Compromise, and Political Influence in Scientific Research.* New York: John Wiley, 1992.

Belmont Report: Ethical Principles and Guidelines for the Protection of Human Subjects of Research. Http://www.ohsr.od.nih.gov/guidelines/belmont.html, accessed October 8, 2005.

Bentham, Jeremy. *An Introduction to the Principles of Morals and Legislation.* Oxford: Clarendon, 1996.

"Bioethics: Gene Therapy Business: The Tragic Case of Jesse Gelsinger." *News Weekly*, August 2, 2000.

Boas, Ralph. "The Author and the Editor." In Gerald Alexanderson and Dale Mugler (eds.), *Lion Hunting and Other Mathematical Pursuits.* Washington, D.C.: American Mathematical Association, 1996.

Bohlin, Ray. "The Controversy over Stem Cell Research." Http://www.probe.org/content/view/884/67, accessed October 8, 2005.

Bonhoeffer, Dietrich. *Letters and Papers from Prison.* London: SCM Press, 1967.

Boyd, W. M., et al. "The Chronic Oral Toxicity of Caffeine." *Canadian Journal of Physiology and Pharmocology* 43 (1965): 95ff.

Braybrooke, David. "Ideology." In Paul Edwards (ed.), *Encyclopedia of Philosophy,* vol. 2, p. 126. New York: Simon and Schuster, 1996.

Broad, William, and Nicolas Wade. *Betrayers of the Truth.* New York: Simon and Schuster, 1982.

Brody, Baruch. *The Ethics of Biomedical Research: An International Perspective.* Oxford: Oxford University Press, 1998.

Burne, Jerome. "The Drug Pushers." *Guardian* (UK), April 11, 2002.

Callahan, Daniel. "Death and the Research Imperative." *New England Journal of Medicine* 342 (2000): 654–56.

Carson, Joe. "Striking a Blow for Scientific Integrity." Government Accountability Project (GAP). Http://www.asa3.org/archive/asa/199703/0088.html, accessed October 8, 2005.

CDC Tuskegee Syphilis Study Page. Http://www.cdc.gov/Nchstp/od/tuskegee/index.html, accessed October 8, 2005.

Circumcision Information and Resource Pages. "Pain of Circumcision and Pain Control." Http://www.cirp.org/library/pain, accessed October 8, 2005.

Clark, Ronald. *Einstein: The Life and Times.* New York: Avon, 1994.

CNN/*Time* poll. "Most Americans Say Cloning Is Wrong." March 1, 1997. Http://www.CNN.COM/TECH/9703/01/clone.poll, accessed October 8, 2005.

Cohen, Carl. "The Case for the Use of Animals in Biomedical Research." *New England Journal of Medicine* 315, no. 14 (October 2, 1986): 865–870.

Committee for Research and Ethical Issues of the International Association for the Study of Pain. "Ethical Guidelines for Investigations of Experimental Pain in Conscious Animals." *Pain* 16 (1983): 109–10.

Cook, Janine Denis. "Good Laboratory Practice (GLP) versus CLIA." Guest essay. Http://www.westgard.com/guest16.htm, accessed October 8, 2005.

Cope, Doris K. "Neonatal Pain: The Evolution of an Idea." *American Society of Anesthesiologists Newsletter* 62, no. 9 (September 1998): 6–8.

Craig, K. D. "Implications of Concepts of Consciousness for Understanding Pain Behavior and the Definition of Pain." *Pain Research and Management* 2 (1997): 111–17.

Crawford, U. "Religious Groups Join Animal Patent Battle." *Science* 237 (1987): 480–81.

Crichton, Michael. *Jurassic Park.* New York: Knopf, 1990.

Cuneo, Michael W. *American Exorcism: Expelling Demons in the Land of Plenty.* New York: Broadway Books, 2002.

Davidson, Elizabeth W., Heather Cate, Cecil Lewis, and Melanie Hunter. "Data Manipulation in the Undergraduate Laboratory: What Are We Teaching?" *Proceedings of the First International ORI Research Conference on Research Integrity*, November 2000. Http://ori.dhhs.gov/documents/proceedings_rri.pdf, accessed October 8, 2005.

Duhem, Pierre. *The Aim and Structure of Physical Theory*. Princeton: Princeton University Press, 1954.

Ehrenfeld, David. *The Arrogance of Humanism.* New York: Oxford University Press, 1978.

Engle, S. J., et al. "HPRT-APRT-Deficient Mice Are Not a Model for Lesch-Nyhan Syndrome." *Human Molecular Genetics* 5, no. 10 (1996): 1607ff.

EROWID Experience Vaults. Http://www.erowid.org/experiences/subs/exp_ketamine.shtml, accessed October 8, 2005.

Ferrell, B. R., and M. Rhiner. "High-Tech Comfort: Ethical Issues in Cancer Pain Management for the 1990s." *Journal of Clinical Ethics* 2 (1991): 108–15.

Feyerabend, Paul K. *Against Method.* London: Verso, 1993.

Forman, Paul. "Weimar Culture, Causality and Quantum Theory, 1918–1927: Adaptation by German Physicists and Mathematicians to a Hostile Intellectual Environment." In R. McCormmach (ed.), *Historical Studies in the Physical Sciences.* Philadelphia: University of Pennsylvania Press, 1971.

Fox, Michael. *Superpigs and Wondercorn.* New York: Lyons and Burford, 1992.

Garry, F., et al. "Post-Natal Characteristics of Calves Produced by Nuclear Transfer Cloning." *Theriogenology* 45 (1996): 141–52.

Gaskell, George, et al. "Europe Ambivalent on Biotechnology." *Nature* 387 (1997): 845ff.

Glut, Donald. *The Frankenstein Catalogue.* Jefferson, N.C.: McFarland, 1984.

Goldhagen, Daniel. *Hitler's Willing Executioners.* New York: Knopf, 1996.

Gray, Elizabeth. "Do Respiratory Patients Receive Adequate Analgesia?" *Nursing Standard* 13, no. 47 (August 11, 1999): 32–35.

Hallerman, E. J., and A. R. Kapuscinski. "Transgenic Fish and Public Policy: Anticipating Environmental Impacts of Transgenic Fish." *Fisheries* 15, no. 1 (1990): 2–12.

Hallerman, E. J., and A. R. Kapuscinski. "Transgenic Fish and Public Policy: Patenting Transgenic Fish." *Fisheries* 15, no. 1 (1990): 21–25.

Hallerman, E. J., and A. R. Kapuscinski. "Transgenic Fish and Public Policy: Regulatory Concerns." *Fisheries* 15, no. 1 (1990): 12–21.

Harrison, Ruth. *Animal Machines.* London: Vincent Stuart, 1964.

Hobbes, Thomas. *Leviathan*, ed. J. C. A. Gaskin. Oxford: Oxford University Press, 1998.

Hofstadter, Richard. *Anti-Intellectualism in American Life.* New York: Knopf, 1964.

Hooper, M. R., et al. "HPRT-Deficient (Lesch-Nyhan) Mouse Embryos Derived from Germline Colonization by Cultured Cells." *Nature* 336 (1987): 292–95.

Hume, David. *A Treatise of Human Nature*, ed. L. A. Selby-Bigge. Oxford: Oxford University Press, 1960.

Hunt, Linda. *Secret Agenda: The United States Government, Nazi Scientists and Project Paper Clip, 1945–1990.* New York: St. Martins, 1991.

Illich, Ivan. *Medical Nemesis: The Expropriation of Health.* New York: Pantheon, 1976.

Institute of Laboratory Animal Resources. *Guide for the Care and Use of Laboratory Animals.* Http://www.nap.edu/readingroom/books/labrats, accessed October 8, 2005.

International Association for the Study of Pain. "IASP Pain Terminology." Http://www.iasp-pain.org/terms-p.html, accessed October 8, 2005.

International Association for the Study of Pain. "Pain Terms: A List with Definitions and Notes on Usage Recommended by the IASP Subcommittee on Taxonomy." *Pain* 6 (1979): 249–52.

Jinnah, H. A., et al. "Dopamine Deficiency in a Genetic Mouse Model of Lesch-Nyhan Disease." *Journal of Neuroscience* 14 (1994): 1164ff.

Johnstone, R. E. "A Ketamine Trip." *Anesthesiology* 39, no. 4 (October 1973): 460–61.

Kant, Immanuel. *Foundations of the Metaphysics of Morals,* trans. Lewis White Beck. New York: Liberal Arts Press, 1959.

Karson, Evelyn M. "Principles of Gene Transfer and the Treatment of Disease." In N. First and F. P. Haseltine (eds.), *Transgenic Animals.* Boston: Butterworth-Heinemann, 1991.

Katz, Jay. *Experimentation with Human Beings.* New York: Russell Sage Foundation, 1972.

Katz, Jay. "The Regulation of Human Experimentation in the United States – A Personal Odyssey." *IRB* 9 (1987): 1–6.

Keeton, W. T., and J. L. Gould. *Biological Science.* New York: W. W. Norton, 1986.

Kelley, W. N., and J. B. Wyngaarden. "Clinical Syndromes Associated with Hypoxanthine-Guanine Phosphororibosyltransferase Deficiency." In J. B. Stanbury et al. (eds.), *The Metabolic Basis of Inherited Disease,* 5th ed. New York: McGraw-Hill, 1983, chapter 51.

Kirtz, Bill. "Healthy Skepticism." Http://www.numag.nev.edu/9701/TalkoftheGown.html, accessed October 8, 2005.

Kitchell, Ralph L., and Howard H. Erickson. *Animal Pain: Perception and Alleviation.* Bethesda, Md.: American Physiological Society, 1983.

Kitchell, Ralph, and Michael Guinan. "The Nature of Pain in Animals." In B. E. Rollin and M. L. Kesel (eds.), *The Experimental Animal in Biomedical Research,* vol. 1. Boca Raton: CRC Press, 1990.

Kitchen, H., A. L. Aaronson, J. L. Bittle, et al. "Panel Report on the Colloquium on Recognition and Alleviation of Animal Pain and Distress." *JAVMA* 191, no. 10 (1987): 1186–91.

Kuehn, M. R., et al. "A Potential Model for Lesch-Nyhan Syndrome through Introduction of HPRT Mutations into Mice." *Nature* 326 (1987): 295–98.

Kuhn, Thomas S. *The Structure of Scientific Revolutions,* 3rd ed. Chicago: University of Chicago Press, 1996.

Kulnych, J., and D. Korn. "The Effect of the New Federal Medical Privacy Rule on Research." *New England Journal of Medicine* 356 (2002): 201–4.

Laurance, Jeremy. "Licensed to Kill – Facts the Drug Firms Conceal." *New Zealand Herald*, June 3, 2004.

Lebaron, Charles. *Gentle Vengeance: An Account of the First Years at Harvard Medical School.* New York: Richard Marek, 1981.

Lee, B. H. "Managing Pain in the Human Neonates: Implications for Animals." *JAVMA* 221 (2002): 233–37.

Levine, Robert. *Ethics and Regulation of Clinical Research.* New Haven: Yale University Press, 1988.

Lewis, D. B., et al. "Osteoporosis Induced in Mice by Overproduction of Interleukin-4." *Proceedings of the National Academy of Sciences* 90, no. 24 (1993): 11618–22.

Liebeskind, J., and R. Melzack. "Meeting a Need for Education in Pain Management." *Pain* 30 (1987): 1–2.

Lifton, Robert Jay. *The Nazi Doctors: Medical Killing and the Psychology of Genocide.* New York: Basic Books, 1986.

Lumb, W., and E. W. Jones. *Veterinary Anesthesia.* Philadelphia: Lea and Febiger, 1973.

Mader, Sylvia. *Biology: Evolution, Diversity, and the Environment.* Dubuque, Iowa: W. M. C. Brown, 1987.

Marks, R. M., and E. J. Sacher. "Undertreatment of Medical Inpatients with Narcotic Analgesics." *Animals of Internal Medicine* 78 (1973): 173–81.

Mayr, Ernst. "The Species Concept: Semantics versus Semantics." *Evolution* 3 (1949): 371–72.

McNeish, J. D., et al. "Legless, a Novel Mutation Found in PHT1-1 Transgenic Mice." *Science* 241 (1988): 837–39.

Merskey, H. "Pain, Consciousness and Behavior." *Pain Research and Management* 2 (1997): 118–22.

Merskey, H. "Response to Editorial: New Perspectives on the Definition of Pain." *Pain* 60 (1996): 209.

Merskey, H., and N. Bogduk. *Classification of Chronic Pain: Description of Chronic Pain Syndromes and Definition of Pain Terms.* Seattle: ASP Press, 1994.

MetroHealth Anesthesia. "History of Ketamine." Http://www.metrohealthanesthesia.com/edu/ivanes/ketamine1.htm, accessed October 8, 2005.

Milgram, Stanley. *Obedience to Authority: An Experimental View.* New York: Harper and Row, 1975.

Mill, John Stuart. *Utilitarianism.* Indianapolis: Hackett, 1979.

Moreno, Jonathan. *Undue Risk: Secret State Experiments on Humans.* New York: W. H. Freeman, 2000.

National Bioethics Advisory Commission. *Cloning Human Beings.* 1997. Http://www.georgetown.edu/research/nrcbl/nbac/pubs.html, accessed October 8, 2005.

National Science Foundation. *Science Indicators.* Washington, D.C.: National Science Foundation, 1988.

Nolan, Kevin. "Guidelines on Good Research Practice." October 23, 2003. The Wellcome Trust. Http://www.rcsi.ie/research/Education_and_Outreach/Postdoctoral_Seminars/Goodresearchpractice.ppt, accessed October 8, 2005.

Nuremberg Code. Http://ohsr.od.nih.gov/nuremberg.php.

O'Donoghue, P. "European Regulation of Animal Experiments." *Lab Animal* (September 1991): 20ff.

Office of Technology Assessment. *New Developments in Biotechnology: Public Perceptions of Biotechnology.* Washington, D.C.: Office of Technology Assessment, 1987.

Pappworth, M. H. *Human Guinea Pigs: Experimentation on Man.* Boston: Beacon, 1968.

Pernick, Martin. *A Calculus of Suffering: Pain, Professionalism and Anesthesia in Nineteenth Century America.* New York: Columbia University Press, 1985.

Phinney, Carolyn. "Toward Some Scientific Objectivity in the Investigation of Scientific Misconduct." Presented at the DPR Symposium "Whistleblowers, Advocates and the Law," held at the National ACS Meeting, New York City, August 27, 1991. Http://academic.mu.edu/phys/buxtoni/DW/Gallo/scientific_misconduct.htm, accessed October 8, 2005.

Plato. *Protagoras and Meno,* trans. W. K. C. Guthrie. Baltimore: Penguin, 1961.

Plato. *The Republic of Plato,* translated with introduction and notes by F. M. Cornford. Oxford: Clarendon, 1941.

Pope John Paul II. "Address to President Bush." Http://www.americancatholic.org/news/stemcell/pope_to_bush.asp, accessed October 8, 2005.

Pursel, Vernon, et al. "Genetic Engineering of Livestock." *Science* 244 (1989): 1281–88.

Ramey, David W., and Bernard E. Rollin. *Complementary and Alternative Veterinary Medicine Considered.* Ames, Iowa: Blackwell, 2004.

Rawls, John. *A Theory of Justice.* Cambridge, Mass.: Belknap Press, 1971.

Redhead, N. J., et al. "Mice with Adenine Phosphororibosyltransferase Deficiency Develop Fatal 2,8-Dihydroxyadenine Lithiasis." *Human Gene Therapy* 7, no. 13 (1996): 1491ff.

Reid, Thomas. *Inquiry into the Human Mind on the Principles of Common Sense,* ed. T. Duggan. 1764; Chicago: University of Chicago Press, 1970.

Rifkin, Jeremy. *Declaration of a Heretic.* Boston: Routledge and Kegan Paul, 1985.

Riley, Vernon. "Mouse Mammary Tumors: Alteration of Incidence as Apparent Function of Stress." *Science* 189 (1975): 465–69.

Robins, L., et al. "Drug Use in U.S. Army Enlisted Men in Vietnam: A Follow-up on Their Return Home." *American Journal of Epidemology* 99, no. 4 (1974): 235–49.

Rollin, Bernard E. *Animal Rights and Human Morality,* 2nd ed. Buffalo, N.Y.: Prometheus, 1992.

Rollin, Bernard E. "Ethics, Animal Welfare, and ACUCs." In John P. Gluck, Tony DiPasquale, and F. Barbara Orlans (eds.), *Applied Ethics in Animal Research: Philosophy, Regulation, and Laboratory Applications.* West Lafayette, Ind.: Purdue University Press, 2002.

Rollin, Bernard E. "Ethics, Science, and Antimicrobial Resistance." *Journal of Agricultural and Environmental Ethics* 14, no. 1 (2001): 29–37.
Rollin, Bernard E. *Farm Animal Welfare.* Ames: Iowa State University Press, 1993.
Rollin, Bernard E. *The Frankenstein Syndrome: Ethical and Social Issues in the Genetic Engineering of Animals.* New York: Cambridge University Press, 1995.
Rollin, Bernard E. "The Frankenstein Thing." In J. W. Evans and A. Hollaender (eds.), *Genetic Engineering of Animals: An Agricultural Perspective.* New York: Plenum, 1986.
Rollin, Bernard E. "Pain, Paradox, and Value." *Bioethics* 3, no. 3 (July 1989): 211–26.
Rollin, Bernard E. "Some Conceptual and Ethical Concerns about Current Views of Pain." *Pain Forum* 8, no. 2 (1999): 78–83.
Rollin, Bernard E. "There Is Only One Categorical Imperative." *Kant-Studien* 67, no. 1 (1976): 60–72.
Rollin, Bernard E. *The Unheeded Cry: Animal Consciousness, Animal Pain, and Science.* Oxford: Oxford University Press, 1989; revised ed., Ames: Iowa State University Press, 1998.
Rollin, Bernard E. "Updating Veterinary Ethics." *JAVMA* 173 (1978): 1015–18.
Rollin, Bernard E., and M. Lynne Kesel. *The Experimental Animal in Biomedical Research,* vols. 1 and 2. Boca Raton: CRC Press, 1990, 1995.
Rolston, Holmes. "Duties to Endangered Species." *Bioscience* 35 (1995): 718–26.
Rolston, Holmes. *Environmental Ethics: Duties to and Values in the Natural World.* Philadelphia: Temple University Press, 1988.
Rolston, Holmes. *Philosophy Gone Wild.* Buffalo, N.Y.: Prometheus, 1989.
Rosenhan, David, and Martin Seligman. *Abnormal Psychology.* New York: Norton, 1984.
Rosenthal, Robert. *Experimental Effects in Behavioral Research.* New York: Appleton, 1966.
Rowan, Andrew. "Alternatives to Animals: Philosophy, History, and Practice." In B. Rollin and M. L. Kesel (eds.), *The Experimental Animal in Biomedical Research,* vol. 2. Boca Raton: CRC Press, 1989.
Ruse, Michael. *The Philosophy of Biology.* London: Hutchinson University Library, 1973.
Russell, Bertrand. *The Problems of Philosophy.* Oxford: Oxford University Press, 1951.
Ryan, Kenneth S. "Ryan Commission Report: Integrity and Misconduct in Research." Http://www.faseb.org/opra/cri.html, accessed October 8, 2005.
Sacks, Oliver. *Awakenings.* New York: Harper, 1990.
Sample, Ian. "The Giants of Science." *Guardian,* September 25, 2003. Http://www.buzzle.com/editorials/9-25-2003-45849-asp, accessed October 8, 2005.
Schilpp, Paul Arthur, ed. *Albert Einstein: Philosopher/Scientist.* New York: Tudor, 1951.
Scriver, C. R., et al. *The Metabolic Basis of Inherited Disease,* vols. 1 and 2, 6th ed. New York: McGraw Hill, 1989.

Secretary's Donation Initiative. Http://www.organdonor.gov/secinitiative.htm.

Shabecoff, Phillip. "Head of the EPA Bars Nazi Data in Study on Gas." *New York Times*, March 23, 1988, p. 1.

Shafir, Sharoni, and Donald Kennedy. "Research Misconduct: Media Exaggerates Results of a Survey." *The Scientist* 12 [13], June 22, 1998.

Shalev, Moshe. "European and U.S. Regulation of Monoclonal Antibodies." *Lab Animal* (January 1998): 15ff.

Sharmann, W. "Improved Housing of Mice, Rats, and Guinea Pigs: A Contribution to the Refinement of Animal Experiments." *Alternatives to Laboratory Animals* 19, no. 1 (February 1991): 108–14.

Smith, Richard. "Research Misconduct and Biomedical Journals." Http://bmj.bmjjournals.com/talks/physics2/tsld001.htm, accessed October 8, 2005.

Spinoza, Baruch. *Ethics*. London: Oxford University Press, 1923.

Statement of 24 Religious Leaders against Animal Patenting. Distributed by the Foundation for Economic Trends, 1987.

Stewart, Walter. Walter Stewart's Web site, Http://www.home.t-online.de/home/Bernard.Hiller/wstewart/main.html, accessed summer 2004.

Stolberg, Sheryl Gay. "The Biotech Death of Jesse Gelsinger." *New York Times Magazine*, November 28, 1999, pp. 136–50.

Stout, J. T., and C. T. Caskey. "Hypoxanthine Phosphororibosyltransferase Deficiency: The Lesch-Nyhan Syndrome and Gouty Arthritis." In C. R. Scriver et al. (eds.), *The Metabolic Basis of Inherited Disease*, vol. 1. New York: McGraw-Hill, 1989, chapter 38.

Syphilis Study Legacy Committee Final Report of May 20, 1996. Http://www.med.virginia.edu/hs_library/historical/apology/report.html, accessed October 8, 2005.

te Velde, E. R., et al. "Concerns about Assisted Reproduction." *Lancet* 351 (1998): 1524–25.

Tiedje, J. M., et al. "The Planned Introduction of Genetically Engineered Organisms: Ecological Considerations and Recommendations." *Ecology* 70, no. 2 (1989): 297–315.

Voss, David. "Whatever Happened to Cold Fusion?" *Physics World* (March 1999). Http://physicsweb.org/article/world/12/3/8, accessed October 8, 2005.

Walco, G. R. C. Cassidy, and N. L. Schechter. "Pain, Hurt, and Harm: The Ethics of Pain Control in Infants and Children." *New England Journal of Medicine* 331, no. 8 (1994): 541–44.

"The Willowbrook Letters." From the Lancet, reprinted in Ronald Munson (ed.), *Intervention and Reflection: Basic Issues in Medical Ethics*, 6th ed. Belmont, Calif.: Wadsworth/Thomson, 2000.

Woolfrey, J., and C. Campbell, eds. *Reflections*, special edition: "Human Cloning: Fact, Fiction and Faith." Corvallis: Department of Philosophy, Oregon State University, May 1997.

Index

AALAS: animal research legislation and, 21–22; ignoring ethics by, 111, 119
ACUC legislation: animal activists' criticism of, 124; congressional hearings on, 116–17; development of, 113–20; effectiveness of, 121–28, 243; goals of, 113–15; guide for, 119–20; provisions of, 120–21; scientific ideology erosion with, 121–28; transgenic animals/disease models and, 182
adult stem cells, 210
agriculture: business model of, 191–92; harmful changes in, 4–5; lack of laws regulating research in, 127, 170–71; monoculture in, 165, 192–93. *See also* animal husbandry; factory farming
Akaka, Abraham, 200, 201
Albert Einstein: Philosopher/Scientist, 39
alcoholism as "disease," 25
alternative medicine, 4, 5–6, 244–45
American Association for Laboratory Animal Science. *See* AALAS
American Journal of Obstetrics and Gynecology, 75
American Medical Journal, 255
American Physiological Society, 115, 116, 118
American Psychological Association, 252
American Veterinary Medical Association Code of Ethics, 43
American Veterinary Medical Association conference report on animal pain, 118–19
analgesics, 120, 222
Anand, K. J. S., 223–25, 233
anesthesia definition/description, 222
Animal and Plant Health Inspection Services, 117
Animal Care and Use Committees. *See* ACUC legislation
animal consciousness, 240–41
animal husbandry, 168, 203–05
Animal Machines (Harrison), 109
animal nature/*telos*, 108
animal pain: animal stoicism and, 242; AVMA conference report on, 118–19; companion animals and, 243; human pain models and, 118, 119, 236, 241; as more acute, 236; paralytic drugs and, 110; perception and, 51; researchers early dealing with, 117–20; Rimadyl and, 243; scientists' denial of, 28–29, 99, 110, 215–16, 239–43. *See also* ACUC legislation; pain
Animal Pain: Perception and Alleviation, 118, 240
animal research: abuse scandals, 115; cosmetic testing and, 103, 104, 172;

animal research (*cont.*)
"euthanizing" animals, 18; lack of ethics effects, 109–10, 122, 124; medical schools and, 18–19; middle ground on, 100, 113; overview, 18–20; paralytic drugs and, 121; proper environment for animals, 109–10, 115, 121, 172, 173; scientists' emotional defense of, 21–22; scientists' inability to defend ethically, 110–13; social ethics and, 43–44. *See also* biomedicine; veterinary medicine; veterinary school
animal research legislation: AALAS and, 21–22; scientists' opposition to, 21–22. *See also* ACUC legislation
animal rights, 101, 104, 108
Animal Rights and Human Morality (Rollin), 101
animal welfare: communications to Congress about, 101–02; companion animals and, 104; cruelty to animals, 104, 105–06; factory farming and, 102, 103, 108–09, 168, 172, 203–05; in Europe, 102, 103, 104, 108–09; legislation (U.S.) and, 102–03; social consensus ethics and, 35; society's increasing concern for, 100, 101–06, 108–09, 171–72; sow stalls and, 102; U.S. vs. Europe, 108–09; veal industry and, 102; wildlife management and, 103–04; xenotransplantation and, 209–10. *See also* ACUC legislation; BST controversy
Animal Welfare Act amendment, 117
Annals of Internal Medicine, 251
anti-intellectualism, 3, 109
Anti-Intellectualism in American Life (Hofstadter), 109
APHIS and ACUC, 117
Aristotle: assumptions and, 38; biology vs. physics, 27, 29; *Metaphysics*, 15; nominalism and, 142; *physikoi* and, 15; reductionism and, 139; the "subjective" approach of, 216, 217
Arrogance of Humanism (Ehrenfeld), 190
Asilomar Conference, 160
assumptions: all disciplines and, 38; philosophy and, 37–39; Plato and, 39;
AVMA Code of Ethics, 43
AVMA conference report on animal pain, 118–19
Awakenings (Sacks), 24, 141
Ayer, A. J., 16

Bacon, Francis, 16
Beecher, Henry, 72–73
behaviorism, 27, 217, 218
Bell, Robert, 256, 269
Belmont report/principles, 87–88
Beltsville pig, 168
"beneficence" in Belmont report/principles, 88
Bentham, Jeremy, 58
Berg, Alan, 107
Bergson's life force, 16
Berkeley, George, 237
Betrayers of the Truth (Broad and Wade), 248–49
biblical fundamentalism as ideology, 12
Biological Science (Keeton and Gould), 17
biomedicine: effects of private research on, 255–56; funding and values, 23; humans as models in, 22–23; public's loss of faith in, 4; safety of animal drugs vs. human drugs, 23. *See also* animal research; biotechnology; veterinary medicine; veterinary schools
biotechnology: ACUC legislation and, 127; computer-age comparison, 130–31; Europe's rejection of, 104; Frankenstein myth and, 133–35, 143, 205; human growth hormone gene and, 136–37, 153–54, 168–69, 203; polls on, 130, 131–33, 135–36, 177, 185–86; problems overview, 9–10; scientists' attitudes and public response, 129–33, 190–91, 214. *See also specific technologies*
biotechnology and suffering animals: euthanasia and, 170, 172; lessons to be learned about, 169–70; overview/examples, 168–72
biotechnology as "dangerous": AIDS research example, 157–58; animal disease models and, 166; antibiotics in animal feed example, 160–61; BST controversy, 103, 161–63; chaos theory and, 158–59; common sense vs. science, 159–60; environmental

problems with, 166; ethics of acceptable risk, 160–61; food animals and, 164; Frankenstein myth and, 155–59; *Jurassic Park* example, 158–59; loss of genetic diversity with, 165; Newtonian predictability and, 158–59; overview of dangers, 163–67; pathogen changes/resistance and, 165–66; pleiotropic effects, 163–64; socioeconomic problems, 167; terrorist/military dangers, 166–67; traditional breeding problems example, 163–64
biotechnology as "intrinsically wrong": environmental philosophy and, 149–53; Frankenstein myth and, 133–35; human genetic disease research, 154; mixing of human/animal traits, 153–54; pervasiveness of position, 135–36; public view of, 98, 131–33; reductionist approach and, 138–42; species inviolability and, 143–49; theology and, 136–38
Blake, William, 3
Boas, Roger, 266
Body Shop, 103
Bohlin, Ray, 212
Bonhoeffer, Dietrich, 137
"Boy with Two Fathers," 51
Boys from Brazil, The, 194
Braunwald, Eugene, 251
Brave New World and utilitarianism, 58, 62
Braybrooke, David, 11, 13
British Journal of Obstetrics and Gynecology, 257
Broad, William, 248, 250, 252, 260, 265, 268
Brockovich, Erin, 268
BST controversy, 103, 161–63
bullfighting, 104
Bush, George W., 211

capital punishment, 20
capitalist ideology, 12
careerism and fraud in science, 250–54
Carson, Rachel, 149
Categorical Imperative, 59
cattle, double muscling engineering, 170. *See also* BST controversy
Chamberlain, Geoffrey, 257
chaos theory, 158–59
child abuse as "disease," 25
Chomsky, Noam, 239
circuses, 103
civil rights example of social ethics criticism, 41–42
Clinton, Bill, 42, 178; Bioethics Commission, 130
clitorectomies, 12–13
cloning: of companion animals, 205; exploitation and, 188–89; genetic diversity loss with, 191–94; genetically vs. literally identical individuals concept, 186, 194–96; as "intrinsically wrong," 188–91; "playing God" argument, 208; possible harm to clones, 202–08; public reaction to Dolly, 130, 131, 185–87; religious arguments about, 186–90; risks with cloning animals, 191–94
cloning humans: benefits of, 200, 201–02; "eugenics" and, 199; "family values" and, 196–98; fertility drugs ethics vs. 199; for "spare parts," 200–01; Hitlers and, 194, 195–96; "human dignity" and, 189, 198–99; as "intrinsically wrong," 188–89; legislation against, 207–08; public attitude toward, 186–87; racism and, 199–200; "risks" of, 194–201; social reaction to clones, 205–07
cloning organs, 201. *See also* xenotransplantation
CNN/*Time* magazine survey, 185–86
Cohen, Carl, 113
cold fusion scandal, 267
Colorado State University: ethics education at, 270–72; laboratory animal welfare rating, 271
Commission on Research Integrity, 262
commodification of animals, 203. *See also* factory farming
communitarian concerns and Kantianism, 62
consequentialist (teleological) theories, definition/description, 57. *See also specific theories*
contradictions in ethics, 56–57
Craig, K. D., 233

Crichton, Michael, 158–59
Cuyahoga River burning, 3, 149

Daily Mail, 257
Darsee, John, 251–52, 253
De Bakey, Michael, 116, 117
Declaration of Helsinki, 71
democratic societies, ethical system judging and, 50. *See also* Western democratic societies
Democritus, 139
demonic possession/exorcism, 6
Denver Post, 107
deontological theories: definition/description, 57; examples, 59;
Descartes, René, 26, 216, 232, 233, 236, 261
Dewey, John, 138
disease: concept of, 24–26, 140; individual variations in, 23–24, 140–41; valuational components of, 23
distress, 245–46
diversity. *See* multiculturalism
Dole, Bob, 117
Dolly and cloning, 130, 131, 185–87
Driesch's postulation of "entelechies," 16
Duchamp, Marcel, 38
Duhem, Pierre, 261–62

Earth Day, 3, 150
Eddington, Sir Arthur, 258–59
Ehrenfeld, David, 190
Einsatzkommando, 27, 68
Einstein, Albert, 16, 39
electrical transmission line dangers, 157
Ellsberg, Daniel, 268
embryonic stem cell research: abortion views and, 212–13; beginning of human life and, 211–14; benefits of, 210–11; debate over, 210, 211–14; overview, 210–14; therapeutic cloning and, 213–14
embryonic stem cells, 210
"entelechies," 16
entrepreneurship and fraud in science, 256–57
environmental considerations for animals, 109–10, 115, 121, 172, 173
environmental degradation, 3
environmental philosophy, 149–53
environmentalism growth, 3–4
EPA, 26, 69
ethical decisions: contradictions and, 56–57; overview, 54–56. *See also* ethical theories
ethical relativism: arguments against, 46–50; definition/description, 42, 46–47; as self-defeating, 47
ethical rules/edicts: consistency of, 48; definition/description, 45–46; differences in people and, 48; as essential to society, 47–48; types of rules needed, 47–48
ethical theories: individual vs. society, 62–63; moral psychology and, 61–62; religion and, 57; societal differences in, 63; types of, 57. *See also specific theories*
ethics: AALAS meetings and, 111; of acceptable risk, 160–61; changes in, 6–8, 31; dog abuse case, 52–54; ethical judgments, 50–54; $Ethics_1$ vs. $Ethics_2$, 31–32; judging systems, 47–50; legal system and, 33; medical ethics overview, 8–9; objectivity and, 32–33; recognizing ethical issues, 50–54; scientists' education in, 270–74; sexual behavior and, 34–35; social order and, 45–46; social policy and, 33–34; system selection for reincarnation, 49–50. *See also specific types of ethics*
Ethics and Regulation of Clinical Research (Levine), 86
ethics ombudsman, 270–74
$Ethics_1$: components of, 32; definition/description, 31; $Ethics_2$ vs., 31–32, 44–45; monitoring needs of, 37
$Ethics_2$: definition/description, 31, 37; $Ethics_1$ vs., 31–32, 44–45; philosophy and, 37. *See also* philosophy
eugenics and cloning, 199
exorcism/demonic possession, 6
extinction, 144–45, 193–94

factory farming, 102, 103, 108–09, 168, 172, 203–05
Feder, Ned, 268
Felig, Philip, 256–57
Ferrell, B. R., 221
Feyerabend, Paul, 258
Fischer, Ronald, 259
Fleischmann, Martin, 267

Forman, Paul, 30
Fox, Michael A., 113
Frankenstein (Shelley), 2
Frankenstein myth and biotechnology, 133–35, 143, 153, 155–59, 205
Frankenstein Syndrome, The, 133–35
fraud in science: accountability and, 262–63; biomedical research by private companies and, 255–56; careerism and, 250–54; Darsee case, 251–52; effects of, 261–62; entrepreneurship and, 256–57; gene therapy and, 254–55; gift authorship and, 257–58; from laziness, 260; loss of autonomy and, 262–63; money and, 254–57; Neurontin example, 255; peer review and, 263–64; pharmaceutical companies and, 255–56; quantity vs. quality and, 251; referee process and, 264–67; replicability and, 267–68; response to, 260–63; self-policing methods, 262–68; statistics/data on, 252–54; theoretical bias and, 258–60; whistleblowing and, 268–70
fur wearing, 104

Galileo, 26
gambling as "disease," 25
Gaskell, George, 104, 131, 177
Gelsinger, Jesse, 87, 254
Gelsinger, Paul, 255
gene libraries, 165, 193–94
gene therapy, 87, 254–55
genetic diseases: Lesch-Nyhan's disease, 175–78; somatic cell therapy and, 196; statistics on, 174; therapeutic abortion and, 175, 176, 178. *See also* transgenic animals/disease models
Gentle Vengeance, 18
gift authorship, 257–58
Glamour, 104
"God of the gaps," 137
Golden Rule, 48–49
Goldhagen, Daniel, 14–15, 67, 68
Good Laboratory Practices Act, 260
Gorman, Harry: background, 100–01; pain relief for animals, 110; veterinary medicine ethics, 43
Gould, J. L., 17
Grandin, Temple, 203
Green Revolution, 1, 4–5
Gresham, Thomas, 129
Gresham's law, 129; Gresham's law for ethics, 9, 129, 185

"halo" effect, 51
Handler, Philip, 261
happiness in utilitarianism, 58
Harakas, Stanley, 188–89
Harrison, Ruth, 109
Hatch, Orrin, 211
Hickey, P. R., 223–25
Hitler, Adolf, 67
Hitler's Willing Executioners (Goldhagen), 14
Hobbes, Thomas, 45, 47, 49
Hofstadter, Richard, 109
human equality and ideology, 12
Human Guinea Pigs (Pappworth), 73–74, 79
Human Radiation Experiments Web site, 80
human research. *See* research on humans
Human Use of Human Beings, The (Wiener), 131
Hume, David, 118, 160, 247
hunting, 104

IASP: modification of pain definition, 239; pain definition of, 231–34
Icelandic epic poems, 46
ideology: biblical fundamentalism as, 12; definition/description, 11, 13; hatreds and, 14–15; human equality and, 12; Marxism and, 12; of militant Muslims, 13; multiculturalism and, 12–13; Nazis and, 14–15, 67–69; racism and, 13–14; reasons for, 13; research on humans and, 67–69; Ten Commandments' interpretation, 12; thinking and, 13–14. *See also* scientific ideology
Impure Science: Fraud, Compromise, and Political Influence on Scientific Research (Bell), 256, 269
individual rights in democracies, 63–64
informed consent: comprehension of information, 89–90; information with, 88–89; statement of, 88–91; voluntariness and, 90–91
Ingelfinger, Franz, 86
Inquisition, 67

Institutional Review Board, 82
Institutional Review Boards (IRBs), 86–97
interdisciplinary teaching, 51–52
International Association for the Study of Pain. *See* IASP
International Conference on Genetic Engineering of Animals, 133
intrinsic value, 60–61, 153

Jacob's Ladder, 80
Jewish Chronic Disease Hospital experiments, 81
Johnson, Lyndon, 41–42
Joint Commission on Accreditation of Healthcare Organizations, 226
Jones, E. W., 119, 240
Journal of the American Medical Association, 256
Journal of Clinical Ethics, 221
"judo vs. sumo" approach, 40–42, 106–07
Jurassic Park (Crichton), 158–59
"justice" in Belmont report/principles, 88

Kamin, Leon, 259
Kant, Immanuel: intrinsic value, 60–61; morality, 48; responsibility, 25; test of universality, 59–60
Kantianism: criticism of, 61–62; overview, 59–61
Katz, Jay, 71, 83, 86
Keeton, W. T., 17
Kelley, W. N., 175, 176
Kennedy, Donald, 116, 261
ketamine: background information on, 227–28; flashbacks/"bad trips" with, 228–31; "illegal" use of, 230; personal account of effects, 228–30; use in elderly/young, 230–31; veterinary medicine use of, 230
killing (humans): logical positivism and, 17, 20; Ten Commandments' interpretation and, 12
King Kong, 168
King, Martin Luther, 32, 39, 40
Kitchell, Ralph, 236
Kitchen, Hiram, 118
Koshland, Daniel, 261
Kuhn, Thomas, 258

Lab Animal, 171
Lancet, 206
Language, Truth, and Logic (Ayer), 16
laziness and fraud in science, 260
Lederer, Susan, 79
Lesch-Nyhan's disease, 175–78
Levine, Robert, 86
Liebeskind, John, 220–21
Lifton, Robert Jay, 67, 68
logical positivism, 16–17, 20
Losonsky, Michael, 203
LSD experiments, 80
Lumb, W., 119, 240

Mader, S., 17
Mahoney, Michael J., 265
Manhattan Project, 20
marijuana use in medicine, 227
Marks, R. M., 219, 221
martial arts example, 40–42, 106–07
Marxist critique of capitalism, 12
mathematics envy, 29
Mayr, Ernst, 146
medical marijuana, 227
medicine: as art and science, 218; pain treatment in, 219–28; physicalization of, 218, 219; scientific ideology and, 83–84
Meditation One (Descartes), 261
Melcher, John, 117
Melzack, Ron, 220–21
Mendel, Gregor, 259
mentally defective/ill and abuse, 76, 81
Merskey, Harold, 232–33
Metabolic Basis of Inherited Disease, The, 174
Metaphysics (Aristotle), 15
Michelson-Morley experiment example, 258
Milgram, Stanley, 14, 96
military (U.S.) research on humans, 80, 85
Mill, John Stuart, 58
money and fraud in science, 254–57
monoclonal antibodies (MAbs) production, 171
monoculture, 192–93
monolithic culture, 13
Monsanto, 10
Moraczewski, Albert, 189
moral arguments vs. religious pronouncements, 187–88

moral principles/Ethics$_1$ principles, 56
moral psychology and ethical theories, 61–62
Moreno, Jonathan, 79
multiculturalism: ethical relativism and, 47; friction/tension and, 13; ideology and, 12–13
Muslim militant ideology, 13
mystics, 5–6

National Bioethics Advisory Commission, 189, 197
National Institutes of Health (NIH): ACUC legislation and, 117, 121; *Guide for the Care and Use of Laboratory Animals*, 115
National Society for Medical Research, 112
Nature, 271
Nazis: experiments on humans by, 26, 66–70; ideology and, 14–15, 67–69;
Neurontin, 255
New England Journal of Medicine, 72, 86, 94, 113, 222, 225
New Scientist, 252
New York Times, 108, 254, 257
Newton, Sir Isaac, 15, 16, 26
Newtonian predictability, 158–59
nominalism/realism debate, 142–43
nuclear power plant disasters, 155, 157
Nuremberg Code: criticism of, 66–71; description/principles of, 70–71; U.S. reception of, 71–72
Nuremburg trial, 70
Nursing Standard, 221

obesity as "disease," 24–25
Of Human Bondage (Maugham), 46
Office for Human-Research Protections, 94
Operation Desert Storm experiments, 80
Operation Paper Clip, 69, 80
Ouspenskaya, Maria, 135

pain: addiction fears and, 226–27; alternative medicine and, 244–45; cancer treatment and, 219, 220; children/infants and, 220–21, 222–26, 230–31; effects of IASP's definition of pain, 231–32, 233–34, 239; elderly and, 220–21, 230–31; historical treatment/view of, 221; IASP's definitions of, 231–34, 239; linguistic competence and, 231, 236–39; management change in animals vs. humans, 243–44; medical treatment of, 219–28; neonate surgery and, 222, 223–25; over-the-counter medications and, 244; paper on neonate pain, 223–25; people seeking euthanasia and, 219; physicalization of medicine and, 218, 219; reasons for ignoring in neonates, 223; recognition of, 234–36; scientists' denial of, 215–16; undertreatment of, 219–26. *See also* animal pain
pain behavior, 233
pain control: denial of animal pain and, 28–29, 99, 110, 215–16, 239–43; in animals vs. humans, 243–44; increased interest in (for animals), 243; need for, 219–26; over-the-counter medications and, 244. *See also* ACUC legislation; alternative medicine
Pain, 220–21
Pain Forum, 231
Pappworth, D. H., 73–74, 79
paralytic drugs: definition/description, 222; in neonate open heart surgery, 222; in veterinary medicine, 240
pathogen changes/resistance: antibiotic use and, 165–66; biotechnology and, 165–66
pathogen release experiments, 80
Pearce, Malcolm, 257
peer review, 263–64
People for the Ethical Treatment of Animals (PETA), 54, 102
perception and ethical judgment, 50–54
Pernick, Martin, 221
personal ethics: criticism of, 42–43; definition/description, 34; logical inconsistencies in, 42–43; relationship with social consensus ethics, 34–35
pharmaceutical companies' research, 255–56
philosophy: assumptions and, 37–39; as "not scientific," 30. *See also* Ethics$_2$
Phinney, C., 252–53
physikoi, 15
plagiarism: peer review and, 263; referee process and, 264

Plato: ethical assumptions and, 39; ethical rules, 45, 46; ethical theory, 58; ethics and public policy, 33; ethics' importance, 273; human potential, 45; on social ethics criticism, 39; recollection, 106; selecting ethical system for reincarnation, 49; the sensory and, 27; as utilitarian, 61;
"playing God" argument, 208
pleiotropy, 163–64
pollution, 3
Pons, Stanley, 267
Pope John Paul II, on cloning, 186; faith vs. secular ethics, 188; stem cell research, 211
Principle of Conservation of Welfare, 171
professional ethics: legislative constraints and, 36–37; overview, 35–36, 181; transgenic animals/disease models and, 182–84
professional ethics criticism: loss of autonomy and, 43–44, 181–82; overview, 43–44;
Prohibition: social ethics criticism and, 42; social policy-morality relationship and, 33

quantum theory, 30

racial hygiene of Nazis, 67–68
racing (dogs/horses), 104
racism: human cloning and, 199–200; ideology and, 13–14
radiation experiments on humans, 80
Raelians, 207
Ramsey, Paul, 190
Rascher, Sigmond, 69
Rawls, John, 49
Reagan, Ronald. 1
realism/nominalism debate, 142–43
recollection of ethics: increased concern for animals and, 108; IRBs and, 95–96; pain control and, 243–44; Plato and, 106. *See also* "judo vs. sumo" approach; reminding vs. teaching ethics
reductionism, 29, 138–42, 216–18
Reed, Walter, 80
referee process, 264–67
Reflections, 188
Reid, Thomas, 46, 239
reincarnation and ethical systems, 49–50
religious pronouncements vs. moral arguments, 187–88
reminding vs. teaching ethics, 39–42, 106–07, 128
replicability, 267–68
Republic (Plato), 49, 58, 273
Research Modernization Act, 115
research on humans: abuse accounts by Pappworth, 73–79; assessment of risks and benefits, 91–93; Beecher's articles on, 72–73; behavioral research and, 96; children/infants, 74–75; comprehending unethical research, 81–86; Declaration of Helsinki, 71; drug testing, 79; dying/elderly, 76–77; ethics education and, 97–98; as extensions of operations, 77–78; fundamental human decency and, 66–67; ideology and, 67–69; inducement of illness in, 79; "inferiors," 85; informed consent and, 88–91; Jay Katz story, 82–83; medical training and, 83–84; mentally defective/ill, 76, 81; Nazi research, 66–70; nonpatient volunteers/students, 77; patients, 77–78, 81; patients as controls, 78; pregnant women, 75; prisoners, 69, 76, 85; regulation of, 86–97; scientific ideology and, 81–87; subject selection, 93–94; U.S. attitude toward (post-Nazism), 71–72; U.S. military research, 80, 85; voluntary consent and, 66–71. *See also* Belmont report/principles; Nuremberg Code
"respect for persons" in Belmont report/principles, 87–88
reviewer bias, 264–66
Rhiner, M., 221
Riesman, David, 247
Rifkin, Jeremy, 9, 130, 138–39, 141–43
Rimadyl, 243
risks and benefits, assessment of: ethics of, 160–61; nature/scope of, 91–92; statement of, 91–93; as systematic, 92–93
Rissler, Robert, 117
Rockefeller, John D., 80
rodeo example, 40–41
Rollin, Bernard, 101; biotechnology and ethics of risk, 160–61; rodeo ethics,

40–41; solution for transgenic animal disease models, 179–84; on student thinking about ethics, 44–45. *See also* ACUC legislation
Rolston, Holmes, 152
Rosenthal effect, 50
Rowan, Andrew, 264
Ruse, Michael, 146
Russell, Bertrand, 234
Ryan Commission Report on Integrity and Misconduct in Research, 269, 270, 272

Sacher, E. S., 219, 221
Sacks, Oliver, 24, 140–41
School of Aviation Medicine, 69
Schroeder, Pat, 116–17
science: changing faith in, 1–10; epistemic basis, 22; ethical education and, 270–74; as "ethics/value-free," 17–22, 27–29, 248; humanness of scientists, 247–50; public confidence and, 5, 6. *See also* fraud in science
Science, 110, 261
scientific ideology: ACUC legislation effects on, 121–28; behaviorism and, 27; common sense and, 11, 85, 159–60; as "Common Sense of Science," 28; epistemic basis of science and, 22; ethical issues' recognition and, 54; ethical judgments and, 45–46; ethics education and, 97–98; evolution of, 15; experience and, 15–17; issue of "killing," 17, 20; logical positivism and, 16–17; Manhattan Project and, 20; medical training and, 83–84; public credibility and, 98; reductionist approach in, 29, 216–18; research on humans and, 81–87; science as "ahistorical," 29–30; science as "aphilosophical," 30; value changes and, 26–27. *See also* subjective experience and science
scientific revolution, 26–27, 217
Secundy, Marian Gray, 200
Seed, Richard, 205, 207
Shafir, Sharoni, 261
Shelley, Mary, 3
Shock Society and *Shock* journal, 266–67
Singular, Steve, 107
60 Minutes, 76
social consensus ethic: animal treatment and, 35, 180–81; definition/description, 34; individual rights' priority in, 95; relationship with personal ethics, 34–35; social order and, 34, 45, 180–81
social ethics criticism: civil rights example of, 41–42; martial arts example, 40–42, 106–07; Plato on, 39; Prohibition and, 42; "reminding" vs. teaching, 39–42, 106–07, 128; rodeo example, 40–41
Society for the Study of Pain, 127
Socrates, 37
solipsism, 234
Soman, Vijay, 256–57
somatic cell therapy, 196
Sophists, 46
sow stalls, 102
space shuttle tragedies, 155, 157
species concept, 147
species inviolability: biblical source of view, 145; biotechnology and, 143–49; extinction and, 144–45; scientific community and, 145–47; scientific debate on, 143; species changes and, 147–49
Spinoza, 238
Spira, Henry, 109, 113
stem cell research. *See* embryonic stem cell research
Stevenson, Adlai, 3, 109
Stewart, Walter, 268
Strughold, Hubertus, 69
subject selection, 93–94
Subjected to Science: Human Experimentation in America before the Second World War (Lederer), 79
subjective experience and science, 27, 32–33, 54, 99, 115, 120, 215–19, 220, 223, 225, 227. *See also* scientific ideology
"sumo vs. judo" approach, 40–42, 106–07
syphilis study, Tuskegee, 81

Taliban culture, 13
te Velde, E. R., 206
teaching vs. reminding ethics, 39–42, 106–07, 128
teleological ethical theories. *See* consequentialist (teleological) theories

terrorism and biotechnology, 166–67
Tesco (British) supermarket chain, 103
theoretical bias and fraud in science, 258–60
Theory of Justice (Rawls), 49
therapeutic abortion, 175, 176, 178
therapeutic cloning, 213–14
Thomas, Lewis, 261
Thomson, James, 210
Time, 135
toxicology: nonmammalian model for, 263–64; research in, 127
transgenic animals/disease models: ACUC and, 182; ethical solution for, 179–84; ethics overview, 172–75; Lesch-Nyhan's disease, 175–78; professional ethics and, 182–84; public attitudes on, 178, 180; seriousness of issue, 178–79
truth telling as essential, 46, 48
tsar baalay chaim, 105
Tuskegee syphilis study, 81

Undue Risk: Secret State Experiments on Humans (Moreno), 79
universality test, 59–60
USDA: ACUC legislation and, 121; "distress" and, 122
utilitarianism: criticism of, 58, 61–62; overview, 57–59

values change in scientific revolution, 26–27, 217
Varmus, Harold, 254
veal industry, 102
Veterinary Anesthesia (Lumb and Jones), 119, 239–40
veterinary medicine: denial of animal pain in, 240; ethics vs. etiquette, 100–01; extralabel drug use in, 36, 43–44; ketamine use in, 230; professional ethics and, 43; social fear of irresponsible drug use in, 36, 43–44
veterinary schools: animal treatment in, 18, 19; ethics course for, 100–01; multiple survival surgeries, 19, 106–08, 112
virtue ethics, 61
Von Braun, Wernher, 69

Wade, Nicholas, 248, 250, 252, 260, 265, 268
Walco, G. A., 225–27
Walgren, Doug, 117
Watson, J. B., 217, 218
Waxman, Henry, 116, 117
Western democratic societies: individual rights in, 63–64; individuals vs. society in, 63–64; social ethics of, 62. *See also* democratic societies
"Whistleblower's Bill of Rights," 269
whistleblowing, 268–70
Wiener, Norbert, 130
wildlife management, 103–04
Will I Be All Right, Doctor? (film), 21, 112
Willowbrook State School experiments, 81
Wittgenstein, Ludwig, 17, 55
Wolff, Robert Paul, 84
Wolfle, Tom, 172
World Health Organization: antibiotics in animal feed, 160; health definition, 24, 140
World Medical Association, 71
Wyngaarden, J. B., 19–20, 175, 176
Wyoming's monolithic culture, 13

xenotransplantation: animal welfare and, 209–10; definition/description, 154; overview, 208–10; xenozoonoses and, 208
xenozoonoses and xenotransplantation, 208

yellow fever experiments on humans, 80

zoos, 104